REVISE EDEXCEL AS/A LEVEL
Chemistry

D0715410

REVISION WORKBOOK

Series Consultant: Harry Smith

Author: Nigel Saunders

A note from the publisher

In order to ensure that this resource offers high-quality support for the associated Pearson qualification, it has been through a review process by the awarding body. This process confirms that this resource fully covers the teaching and learning content of the specification or part of a specification at which it is aimed. It also confirms that it demonstrates an appropriate balance between the development of subject skills, knowledge and understanding, in addition to preparation for assessment.

Endorsement does not cover any guidance on assessment activities or processes (e.g. practice questions or advice on how to answer assessment questions) included in the resource nor does it prescribe any particular approach to the teaching or delivery of a related course.

While the publishers have made every attempt to ensure that advice on the qualification and its assessment is accurate, the official specification and associated assessment guidance materials are the only authoritative sources of information and should always be referred to for definitive guidance.

Pearson examiners have not contributed to any sections in this resource relevant to examination papers for which they have responsibility.

Examiners will not use endorsed resources as a source of material for any assessment set by Pearson.

Endorsement of a resource does not mean that the resource is required to achieve this Pearson qualification, nor does it mean that it is the only suitable material available to support the qualification, and any resource lists produced by the awarding body shall include this and other appropriate resources.

For the full range of Pearson revision titles across GCSE, BTEC and AS/A Level visit:
www.pearsonschools.co.uk/revise

ALWAYS LEARNING   PEARSON

Contents

A small bit of small print
Edexcel publishes Sample Assessment Material and the
Specification on its website. This is the official content and
this book should be used in conjunction with it. The questions
in 'Now try this' have been written to help you practise every
topic in the book. Remember: the real exam questions may
not look like this.

Atomic structure and isotopes

1 Complete this table to show the properties of the particles present in an atom.

	Relative mass	Relative charge
Proton	1	1
Neutron	1	0
Electron	$\frac{1}{2000}$	-1

(3 marks)

2 The nucleus of a $^{19}_{9}F$ atom contains:

☐ **A** 19 neutrons and 9 protons

☑ **B** 9 protons and 10 neutrons

☐ **C** 9 protons and 19 neutrons

☐ **D** 10 neutrons and 9 electrons

(1 mark)

3 (a) Define the term **relative atomic mass**.

the weighted mean average mass of an atom of an element compared to $\frac{1}{12}$ of the mass of an atom of carbon-12

(2 marks)

> **Guided**

(b) In terms of the particles present in an atom, explain what is meant by the term **isotopes**.

Isotopes are atoms that have the same number of ...... protons + electrons but different numbers of neutrons ...... **(2 marks)**

4 An atom of element **Z** contains three more neutrons, and two more protons, than an atom of ^{28}Si. State the atomic number and mass number of element **Z**, and its name.

Atomic number ...... 16 ......

Mass number ...... 29 ......

Name of element ...... Sulfur ......

> Use the Periodic Table to help you identify element **Z** from its atomic number.

(3 marks)

$\begin{array}{c} 2\ 9 \\ 1\ 6 \end{array}$ $^{28}_{14}Si$

> **Guided**

5 Which two species have the same number of electrons?

☐ **A** C_2H_4 and O_2

☐ **B** $^{10}_{5}B$ and $^{10}_{4}Be$

☐ **C** Al^{3+} and $_{16}S$

☐ **D** F_2 and ^{18}O

> Option **B** is not correct because the number 10 refers to the number of protons and neutrons in the nucleus of the atoms, not to the number of electrons.

16

(1 mark)

Had a go ☐ Nearly there ☐ Nailed it! ☐

Mass spectrometry

1 Nitrogen has two isotopes, ^{14}N and ^{15}N. One of the peaks in the mass spectrum of nitrogen gas is at *m*/*z* 28, due to $(^{14}N{-}^{14}N)^+$. There are two other peaks in the same region. Identify the species that give rise to these peaks, and state their *m*/*z* values.

....

.... **(2 marks)**

Guided

2 Bromine has two isotopes, with relative isotopic masses of 79 and 81. Four *m*/*z* values are shown below. Which one will occur in a mass spectrum of bromine vapour, Br_2, from an ion with a single positive charge?

☐ ~~A~~ 80

☐ **B** 159

☐ **C** 161

☐ **D** 162

Option **A** is not correct because neither isotope has this relative isotopic mass (it is just the mean of the two values given).

(1 mark)

Guided

3 Chlorine contains 75.78% ^{35}Cl and 24.22% ^{37}Cl.
Calculate the relative atomic mass of chlorine, giving your answer to two decimal places.

[(75.78 × 35) + ... × ....] ÷ 100

= .... **(2 marks)**

4 A sample of nickel contains five isotopes:

Mass/charge ratio	58	60	61	62	64
Relative abundance	68.1	26.2	1.1	3.7	0.9

Calculate its relative atomic mass, giving your answer to one decimal place.

....

.... **(2 marks)**

Maths skills: As in Question 3, you are calculating a weighted mean here. This takes into account the relative abundance of each isotope.

5 Explain how mass spectrometry can be used to find the relative molecular mass of a molecule.

....

.... **(2 marks)**

Shells, sub-shells and orbitals

⟩Guided⟩ **1** Electrons in atoms occupy orbitals. Explain what is meant by the term **orbital**.

An orbital is a region ....

.... **(1 mark)**

2 (a) The electrons in an orbital can be shown as arrows. State the property represented by these arrows.

.... **(1 mark)**

(b) Draw diagrams in the spaces below to show the shapes of an s-orbital and a p-orbital.

s-orbital	**p-orbital**

(2 marks)

3 Complete the table to show the maximum number of electrons that can occupy s-, p- and d-sub-shells.

Sub-shell	s	p	d
Maximum number of electrons			

(1 mark)

4 Which of the following represents the correct maximum numbers of electrons allowed in each of the first four quantum shells?

	1st	2nd	3rd	4th
☑ **A**	2	8	8	18
☐ **B**	2	8	8	32
☐ **C**	2	8	18	32
☐ **D**	2	6	8	18

(1 mark)

5 Which of the following statements correctly describes 1s and 4s sub-shells?

☐ **A** 1s is closer to the nucleus and higher in energy than 4s.

☐ **B** 1s is closer to the nucleus and lower in energy than 4s.

☐ **C** 1s is further from the nucleus and higher in energy than 4s.

☐ **D** 1s is further from the nucleus and lower in energy than 4s. **(1 mark)**

Had a go ☐ Nearly there ☐ Nailed it! ☐

Electronic configurations

Guided

1 (a) Complete the electron configuration for an atom of sodium.

$1s^2\ 2s^2\ 2$.. **(1 mark)**

(b) State the block in the Periodic Table to which sodium belongs.

.... **(1 mark)**

2 Which of the following electronic configurations could be for an atom of an element in Group 5?

☐ **A** $1s^2\ 2s^2\ 2p^1$

☐ **B** $1s^2\ 2s^2\ 2p^6\ 3s^2\ 3p^3$

☐ **C** $1s^2\ 2s^2\ 2p^6\ 3s^2\ 3p^6\ 3d^5$

☐ **D** $1s^2\ 2s^2\ 2p^6\ 3s^2\ 3p^6\ 3d^{10}\ 4s^2\ 4p^5$ **(1 mark)**

3 Which pair of atomic numbers represents two elements in the s-block of the Periodic Table?

☐ **A** 2 and 10

☐ **B** 9 and 17

☐ **C** 3 and 11

☐ **D** 12 and 13 **(1 mark)**

Use the Periodic Table to help you identify two elements in the s-block.

4 (a) Write the electron configuration for an atom of copper.

.... **(1 mark)**

(b) State the block in the Periodic Table to which copper belongs.

.... **(1 mark)**

5 In terms of their electronic configurations, explain why magnesium and calcium have similar chemical properties.

....

.... **(2 marks)**

6 Which of the following diagrams represents the electrons in the ground state of an atom of nitrogen?

	1s	2s	$2p_x$	$2p_y$	$2p_z$
☐ **A**	↑	↑	↑↓	↑↓	↑
☐ **B**	↑↓	↑	↑↓	↑	↑
☐ **C**	↑↓	↑↓	↑↓	↑	
☐ **D**	↑↓	↑↓	↑	↑	↑

(1 mark)

Ionisation energies

1 (a) What is meant by the term **first ionisation energy**?

the amount of energy required to remove one electron from the outermost energy shell of an atom .... **(3 marks)**

(b) Write an equation to represent the reaction accompanying the third ionisation energy for aluminium.

$A^{2+}{}_{(g)} \rightarrow A^{3+} + e^{*-}$ .... **(1 mark)**

2 The electronic configurations for four different elements are shown below. Which one would you expect to have the highest first ionisation energy?

☒ **A** $1s^2\, 2s^2\, 2p^1$

☐ **B** $1s^2\, 2s^2\, 2p^6$

☐ **C** $1s^2\, 2s^2\, 2p^6\, 3s^2\, 3p^1$

☐ **D** $1s^2\, 2s^2\, 2p^6\, 3s^2\, 3p^6$ **(1 mark)**

⟩**Guided**⟩ 3 The table below shows the successive ionisation energies for oxygen.

Electron removed	1	2	3	4	5	6	7	8
Ionisation energy /kJ mol^{-1}	1314	3388	5301	7469	10 989	13 327	71 337	84 080

(a) Explain how you can deduce from these ionisation energies that oxygen is in Group 6.

There is a large jump between the .... 6 & 7 electron removed

.... **(1 mark)**

(b) Explain the trend in these successive ionisation energies.

....

.... **(2 marks)**

4 Explain why first ionisation energies decrease down Group 2.

....

....

.... **(3 marks)**

5 Explain why first ionisation energies generally increase across Period 2 (Li to Ne).

....

....

There are three relevant factors that explain the increase seen.

....

.... **(4 marks)**

Periodicity

1 (a) What is meant by the term **periodicity**?

....

.... **(2 marks)**

(b) Describe the trend in atomic radius across Period 2 (Li to Ne).

.... **(1 mark)**

> **Guided**

2 The table below shows the melting temperatures of the elements in Period 3 (Na to Ar).

Element	Na	Mg	Al	Si	P	S	Cl	Ar
Melting temperature /K	371	922	933	1683	317	392	172	84

(a) Explain, in terms of its structure and bonding, why silicon has a very high melting point.

It has a giant .... structure held together by .... bonds.

A lot of energy is needed to .... **(3 marks)**

(b) Explain, in terms of its structure and bonding, why aluminium has a high melting point.

....

....

.... **(3 marks)**

(c) What can be deduced about the bonds present between atoms of argon from these data?

.... **(1 mark)**

3 The graph shows first ionisation energies for the elements in Period 3 (Na to Ar).

(a) Draw a cross on the graph to show the first ionisation energy of magnesium. **(1 mark)**

(b) Oxygen is directly above sulfur in the Periodic Table. State whether you expect its first ionisation energy to be higher or lower than that of sulfur. Explain your answer.

....

.... **(2 marks)**

Exam skills 1

1 (a) Complete the electronic configurations for:

(i) the phosphorus atom ~~15~~.

$1s^2$ $2s^2$ $2p^6$ $3s^2$ $3p^3$.

(ii) the iron(III) ion, Fe^{3+} ~~2/13~~.

$1s^2$ $2s^2$ $2p^6$ $3s^2$ $3p^6$ $4s^2$ $3d^3$ .... **(2 marks)**

(b) A sample of iron is made up of four isotopes.
The data below were taken from a mass spectrum of this sample.

Mass/charge ratio	54	56	57	58
% abundance	5.85	91.75	2.12	0.28

Calculate the relative atomic mass of the sample, giving your answer to two decimal places.

$\frac{(54 \times 5.85) + (56 \times 91.75) + (57 \times 2.12) + (58 \times 0.28)}{100}$ $= 55.9098$.

(2 marks) -55.9

(c) Write an equation, including state symbols, to show the reaction that happens when the second ionisation energy of iron is measured.

$Fe^+ \rightarrow Fe^{2+} + e^-$ .... **(1 mark)**

(d) (i) Explain the general trend in first ionisation energies for the elements aluminium to argon (Al to Ar) in Period 3.

The first ionisation increases along period 3 from Al to Ar as nuclear
~~increases~~
charge across the period, meaning electrons **(3 marks)**
are more attracted to the nucleus, and so more energy is
same distance + from nucleus / no additional shielding.
(ii) State how sulfur deviates from this trend, and explain your answer. energy is needed to remove an ion.

Ionisation energy of sulfur < ionisation
energy of P (element before), pairing
occurs in ~~3p~~ 3p shell, repulsion **(3 marks)**
occurs, allowing easier ionisation.

Ions

1 Which of the following is the electronic configuration of the aluminium ion, Al^{3+}?

  ☐ **A** $1s^2 2s^2 2p^6$

  ☐ **B** $1s^2 2s^2 2p^6 3s^2 3p^1$

> Use the Periodic Table to identify the atomic number of aluminium and so the number of electrons in its atoms. This lets you work out the number of electrons in its ions.

  ☐ **C** $1s^2 2s^2 2p^6 3s^2$

  ☐ **D** $1s^2 2s^2 2p^6 3s^2 3p^4$

      (1 mark)

2 Which of the following pairs of ions is isoelectronic?

  ☐ **A** Li^+ and F^-

  ☐ **B** N^{3-} and P^{3-}

  ☐ **C** Cl^- and K^+

  ☐ **D** S^{2-} and Ca^+

      (1 mark)

3 Draw electronic configuration diagrams for the following ions. Use dots or crosses in your answers, and include all the electrons and the charges present.

  (a) Mg^{2+} ion   (b) F^- ion.

      (1 mark)   **(1 mark)**

4 Aqueous manganate(VII) ions are purple. Describe what you would expect to see at each electrode during the electrolysis of aqueous potassium manganate(VII), $KMnO_4(aq)$.

  Positive electrode   ....

  Negative electrode   .... **(1 mark)**

5 During the industrial manufacture of copper, copper is purified by electrolysis using a mixture of sulfuric acid and copper(II) sulfate solution. During electrolysis, copper(II) ions are released from the positive electrode and copper atoms are deposited at the negative electrode.

  (a) Explain how copper(II) ions form from copper atoms. Include an equation in your answer.

  ....

  .... **(2 marks)**

  (b) Explain how negatively charged ions form.

  .... **(1 mark)**

Ionic bonds

Guided

1 (a) Describe ionic bonding.

It is the .... of attraction between

.... **(2 marks)**

(b) Suggest a repulsive force that exists in an ionic lattice.

.... **(1 mark)**

2 Which of the following ions has the largest ionic radius?

☐ **A** Na^+

☐ **B** Mg^{2+}

☐ **C** O^{2-}

☐ **D** F^- **(1 mark)**

3 Explain why the aluminium ion, Al^{3+}, has a greater charge density than the sodium ion, Na^+.

....

.... **(2 marks)**

4 Lattice energy is a measure of the strength of ionic bonds. For example, it is −2814 kJ mol⁻¹ for lithium oxide, Li_2O, but only −2232 kJ mol⁻¹ for potassium oxide, K_2O. Explain the difference between the bond strength in these two compounds.

> The factors involved include ionic charge and ionic radius.

....

.... **(2 marks)**

5 Complete this diagram to show the positions of positive and negative ions in solid sodium chloride.

(1 mark)

6 Going down Group 2:

☐ **A** the number of occupied electron shells increases

☐ **B** the atomic radius decreases

☐ **C** the ionic radius decreases

☐ **D** the first ionisation energy increases. **(1 mark)**

Covalent bonds

1 What is a covalent bond?

....

.... **(2 marks)**

2 Lithium aluminium hydride, $LiAlH_4$, contains the ion AlH_4^-. This ion can be represented as shown in the diagram below.

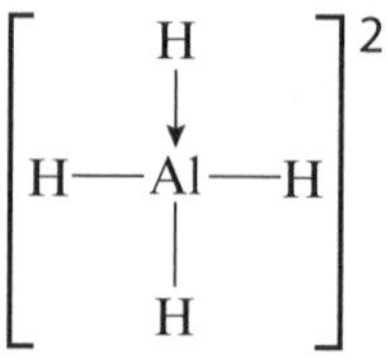

$$\begin{bmatrix} & H & \\ & \downarrow & \\ H & \!\!\!-Al-\!\!\! & H \\ & | & \\ & H & \end{bmatrix}^{2}$$

(a) Name the type of bond represented by Al−H in the diagram.

.... **(1 mark)**

(b) One of the bonds is represented by H→Al in the diagram.

    (i) Name the type of bond this represents.

.... **(1 mark)**

    (ii) Explain, in terms of electrons, how this bond forms.

....

.... **(2 marks)**

3 Draw a dot and cross diagram to show the bonding in the following compounds.

> When two atoms are covalently bonded, show one atom's electrons as dots and the other atom's electrons as crosses. Show the outer shells only unless told otherwise.

(a) Boron trifluoride, BF_3.

(b) Ethane, C_2H_6.

(2 marks)   **(2 marks)**

4 Which of the following atoms has a lone pair of electrons in its outer shell?

  ☐ **A** H in H_2O

  ☐ **B** P in PH_3

  ☐ **C** C in CCl_4

  ☐ **D** H in H_2

(1 mark)

Covalent bond strength

1 Which of the following covalent bonds is the shortest?

    ☐ **A**   F−F

    ☐ **B**   Cl−Cl

    ☐ **C**   Br−Br

    ☐ **D**   I−I   **(1 mark)**

Guided

2 Which of the following statements about carbon–halogen bonds is most likely to be correct?

    ☐ **A**   The C−F bond is shorter and weaker than the C−I bond.

    ☐ **B**   The C−Cl bond is longer and stronger than the C−Br bond.

    ☐ ~~**C**~~   The C−F bond is longer and weaker than the C−I bond.

    ☐ **D**   The C−Br bond is shorter and stronger than the C−I bond.   **(1 mark)**

> Option **C** is not correct because a fluorine atom is much smaller than an iodine atom, so its covalent bond with carbon is shorter.

3 Which of the following bonds is likely to be the strongest?

    ☐ **A**   C−H

    ☐ **B**   C−O

    ☐ **C**   C=O

    ☐ **D**   C≡O   **(1 mark)**

4 Draw dot and cross diagrams to show the bonding in the following molecules.

    (a) Carbon dioxide, CO_2 (b) Ethyne, C_2H_2.

> Carbon dioxide molecules contain double bonds and ethyne molecules contain a triple bond.

           (2 marks)   **(2 marks)**

5 Draw a dot and cross diagram to show the bonding in ammonia aluminium trichloride, which can be represented as $H_3N \rightarrow AlCl_3$.

> The arrow represents a dative covalent bond.

           (3 marks)

Had a go ☐ Nearly there ☐ Nailed it! ☐

Shapes of molecules and ions

1 For each molecule below, draw a diagram to show its shape and bond angle.

(a) CO_2 **(2 marks)**

(b) CH_4 **(2 marks)**

(c) $AlCl_3$ **(2 marks)**

(d) PF_5 **(2 marks)**

2 Which of the following has a shape that is **not** influenced by a lone pair of electrons?

☐ **A** BCl_3

☐ **B** PF_3

☐ **C** CH_3^-

☐ **D** H_2O

(1 mark)

CH_3^- has one more electron than CH_3 would have (and NH_4^+ has one less electron than NH_4 would have).

3 The three species NH_4^+, NH_3 and NH_2^- all contain nitrogen but have different shapes and bond angles. For each one, predict its shape and bond angle, and explain your answer.

(a) NH_4^+ shape and bond angle ....

Explanation ....

.... **(3 marks)**

(b) NH_3 shape and bond angle ....

Explanation ....

.... **(3 marks)**

(c) NH_2^- shape and bond angle ....

Explanation ....

.... **(3 marks)**

4 Which of the following is a correct statement about a molecule of SF_6?

☐ **A** It is trigonal bipyramidal with bond angles of 90° and 120°.

☐ **B** It is trigonal bipyramidal with bond angles of 90° and 180°.

☐ **C** It is octahedral with bond angles of 90° and 120°.

☐ **D** It is octahedral with all bond angles of 90°.

(1 mark)

Electronegativity and bond polarity

Guided

1 Which of the following statements about electronegativity is correct?

☐ ~~A~~ Fluorine is the least electronegative element.

☐ **B** Non-metals have a higher electronegativity than metals.

☐ **C** Electronegativity decreases across Period 3 (Na to Ar).

☐ **D** Electronegativity increases down Group 2.

> Option **A** is not correct because fluorine is the most electronegative element of all.

(1 mark)

2 The table gives the electronegativities of six elements.

Element	C	H	Cl	O	Si
Electronegativity	2.5	2.1	3.0	3.5	1.8

Which of the following compounds would have the greatest ionic character?

☐ **A** H_2O

☐ **B** SiH_4

☐ **C** $SiCl_4$

☐ **D** CCl_4

(1 mark)

3 The C=O bonds in carbon dioxide, CO_2, are polar.

(a) The symbols δ^+ and δ^- may be used to describe the dipole present.

(i) What do these symbols mean?

.... **(1 mark)**

(ii) Which one is used with the oxygen atom? Explain your answer.

....

.... **(2 marks)**

(b) Explain why carbon dioxide is not a polar molecule, even though it contains polar bonds.

....

.... **(2 marks)**

4 Describe how you could carry out an experiment to demonstrate that liquid ethanol is polar.

> State what you would do and the results you would expect to see.

....

.... **(3 marks)**

5 Which of the following molecules is polar?

☐ **A** Boron trichloride, BCl_3

☐ **B** Hexane, C_6H_{14}

☐ **C** Tetrachloromethane, CCl_4

☐ **D** Ammonia, NH_3

(1 mark)

London forces

1 Why do the boiling points of the noble gases increase as you go down the group?

☐ **A** Electronegativity is decreasing, so the covalent bonds between atoms become weaker.

☐ **B** The number of points of contact decreases, so the intermolecular forces become stronger.

☐ **C** The number of electrons increases, so the intermolecular forces are stronger.

☐ **D** Atomic radius increases, so the atoms are further apart.   **(1 mark)**

2 The table describes three structural isomers of C_5H_{12} (molecules that have the same molecular formula but a different arrangement of atoms).

Name	pentane	2-methylbutane	2,2-dimethylpropane
Structure			
Boiling temperature/K	309	301	283

(a) Name the strongest intermolecular force between these molecules, and explain how it arises.

Name of intermolecular force .... **(1 mark)**

Explanation ....

....

.... **(2 marks)**

> **Guided**

(b) Explain why the boiling temperature of pentane is the highest of the three isomers.

Pentane has the greatest ....

.... **(2 marks)**

3 Fractional distillation is used to separate the different components of crude oil. It relies on differences in the boiling temperatures of the alkanes that the oil contains.

(a) State the trend in boiling temperature of the alkanes with increasing chain length.

....

....

For unbranched alkanes, the chain length increases as the M_r increases.

(1 mark)

(b) In terms of the intermolecular forces present, explain your answer to part (a).

....

....

.... **(3 marks)**

Permanent dipoles and hydrogen bonds

1   Hydrogen chloride, HCl, boils at 188 K but fluorine, F_2, boils at 85 K. This is because:

☐ **A** hydrogen chloride has a greater relative formula mass than fluorine

☐ **B** hydrogen chloride contains more electrons than fluorine

☐ **C** hydrogen chloride can form hydrogen bonds but fluorine cannot

☐ **D** permanent dipole–permanent dipole forces exist between hydrogen chloride molecules but not between fluorine molecules.

(1 mark)

2   The graph shows the boiling temperatures of four Group 5 hydrides.

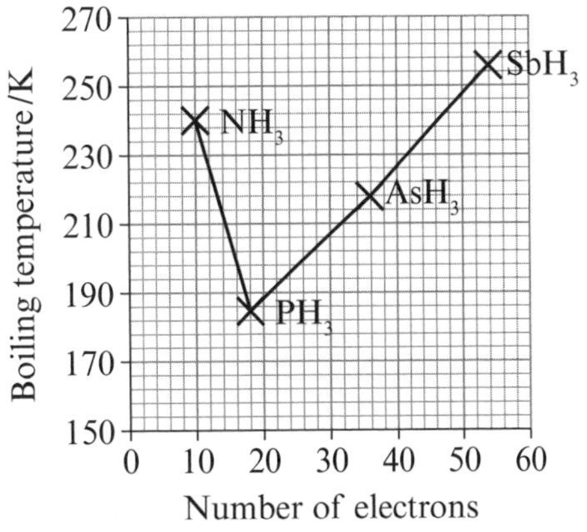

   (a) Explain, in terms of intermolecular forces, the trend in boiling temperature from PH_3 to SbH_3.

   ....
   .... **(2 marks)**

   (b) Predict the boiling temperature of NH_3 due to the forces mentioned in part (a) alone.

   .... **(1 mark)**

   (c) Explain why NH_3 has an unusually high boiling temperature.

   ....
   .... **(3 marks)**

3   (a) Complete this diagram to show a hydrogen bond between two water molecules.

   | Show O–H---O in a straight line. |

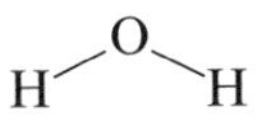

   H⁀O⁀H

   (2 marks)

   (b) Ammonia, NH_3, and water have a similar relative molecular mass.
       Explain why ammonia is a gas at room temperature but water is a liquid.

   ....
   .... **(3 marks)**

Choosing a solvent

1   When an ionic solid such as sodium chloride dissolves in water, individual ions form strong electrostatic attractions with water molecules. This process of hydration is:

☐ **A** endothermic because ion–dipole interactions form

☐ **B** exothermic because ion–dipole interactions form

☐ **C** endothermic for positively charged ions and exothermic for negatively charged ions

☐ **D** exothermic for positively charged ions and endothermic for negatively charged ions.

(1 mark)

2   Ethanol, CH_3CH_2OH, acts as an antiseptic. Ethanol and water are miscible – they mix completely with each other. A 70% mixture of ethanol and water is particularly effective against bacteria.

Complete this diagram to show a hydrogen bond between an ethanol molecule and a water molecule.

> Show O–H---O in a straight line.

$$H_3C-CH_2-O-H$$

(2 marks)

3   The table describes the solubilities in water of three isomers of C_4H_9OH, which are all alcohols.

Alcohol	butan-1-ol	butan-2-ol	2-methylpropan-2-ol
Structure	$H_3C-CH_2-CH_2-CH_2-OH$	$H_3C-CH_2-CH-CH_3$ with OH	$H_3C-C-CH_3$ with CH_3 and OH
Solubility/ $g\,dm^{-3}$ of H_2O	73	290	miscible

(a)   Name the strongest type of intermolecular force you expect there to be between:

(i)   the carbon–carbon chains

.... **(1 mark)**

(ii)   the hydroxyl (−OH) groups

.... **(1 mark)**

(b)   Suggest why 2-methylpropan-2-ol is more soluble in water than the other two alcohols.

....

....

.... **(3 marks)**

(c)   Explain which of the three isomers you expect to be most soluble in cyclohexane, C_6H_{12}.

....

....

.... **(3 marks)**

Giant lattices

1 Which of the following describes the structure and bonding in iodine at room temperature?

☐ A̶ A simple molecular lattice in which iodide ions are attracted by electrostatic forces.

> Option **A** cannot be correct because iodide ions are negatively charged and will repel each other.

☐ **B** A simple molecular lattice in which all iodine atoms are joined together by covalent bonds.

☐ **C** A simple molecular lattice in which iodine molecules are attracted by permanent dipole–permanent dipole forces.

☐ **D** A simple molecular lattice in which iodine molecules are attracted by London forces.

(1 mark)

2 (a) Describe the bonding present in a metal.

.... **(1 mark)**

(b) Describe the structure of a metal.

....

.... **(3 marks)**

3 The diagrams show the structures of diamond and graphite.

Diamond   Graphite

Describe the structure and bonding in:

(a) Diamond ....

....

.... **(3 marks)**

(b) Graphite ....

....

....

.... **(4 marks)**

4 Complete the diagram to show the structure of sodium chloride, NaCl.
One negatively charged ion has already been placed.

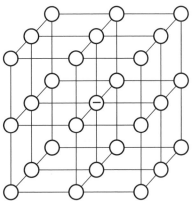

(1 mark)

Structure and properties

1 A substance is solid at room temperature, but it melts at 185 °C. It readily dissolves in water to form a colourless solution that does not conduct electricity. The substance is most likely to:

   ☐ **A** have a giant metallic lattice structure containing delocalised electrons

   ☐ **B** have a giant ionic lattice structure containing oppositely charged ions

   ☐ **C** consist of simple molecules with polar bonds between them

   ☐ **D** have a giant covalent lattice structure.   **(1 mark)**

2 Some substances can conduct electricity. Explain the following observations in terms of the structure of the substance and the types of particle present.

   (a) Aluminium is a good conductor of electricity.

   ....   **(1 mark)**

   (b) Graphite is a good conductor of electricity but diamond is not.

   ....   **(1 mark)**

   (c) Solid sodium chloride does not conduct electricity when   ┌─────────────────┐
   it is solid, but it does conduct when it is molten and   │ Sodium chloride is │
   when it is in aqueous solution.   │ an ionic substance. │
                                                                 └─────────────────┘

   ....

   ....   **(2 marks)**

3 At room temperature, iodine and diamond are both crystalline solids. However, the melting temperature of diamond is very much higher than that of iodine.

   (a) State one similarity in the bonding present in these two substances.

   ....   **(1 mark)**

   (b) State one difference in the structures of these two substances.

   ....   **(1 mark)**

   (c) Explain the observed difference in the melting temperatures of these two substances.

   ....

   ....   **(2 marks)**

4 Uranium hexafluoride, UF_6, is a colourless solid used in the enrichment of uranium for nuclear reactors. It sublimes (turns from a solid to a gas) at 56.5 °C and dissolves in non-polar solvents. Suggest the structure and bonding present in UF_6, giving a reason for your answer.

   ....

   ....

   ....   **(3 marks)**

Exam skills 2

1 (a) Explain why metals have relatively high melting temperatures.

....

....

.... **(3 marks)**

(b) Explain why silica, SiO_2, has a high melting temperature.

....

....

.... **(3 marks)**

(c) Predict the shapes of the following ions. Draw diagrams to show their three-dimensional shapes, predict their bond angles and identify the shapes.

(i) $AlCl_4^-$

Name of shape ....

(ii) ICl_4^-

Name of shape ....   **(6 marks)**

(d) Explain, in terms of intermolecular forces present, why:

(i) hydrogen fluoride, HF, has a higher boiling temperature than hydrogen chloride, HCl

....

(ii) hydrogen iodide, HI, has a higher boiling temperature than hydrogen bromide.

....

....

.... **(6 marks)**

Had a go ☐ Nearly there ☐ Nailed it! ☐

Oxidation numbers

1 State the oxidation number of the element in each of the following species.

(a) O_2 .... **(1 mark)**

(b) S^{2-} .... **(1 mark)**

(c) Al^{3+}.............................. **(1 mark)**

2 What is the oxidation number of phosphorus in P_4O_{10}?

☐ **A** +2.5

☐ **B** +4

☐ **C** +5

☐ **D** +10 **(1 mark)**

3 What is the oxidation number of sulfur in SO_4^{2-}?

☐ **A** +2

☐ **B** +4

☐ **C** +6

☐ **D** +8 **(1 mark)**

4 Chlorine forms various oxides and oxo-anions, including ClO_2, Cl_2O, Cl_2O_6, ClO^-, ClO_3^-, ClO_2^- and ClO_4^-.

> The oxidation number of oxygen is the same in all seven of these species.

(a) In which of these species is the oxidation number for chlorine the lowest? State its value.

Species ....

Oxidation number .... **(2 marks)**

Guided

(b) Name the compound from the list above with the highest oxidation number for chlorine.

> Write a Roman numeral in brackets to represent chlorine's oxidation number.

Chlorine .... oxide **(1 mark)**

5 Write the formula for each of the following compounds.

(a) Copper(I) sulfide .... **(1 mark)**

(b) Copper(II) chloride .... **(1 mark)**

(c) Vanadium(III) fluoride .... **(1 mark)**

(d) Nitrogen(IV) oxide .... **(1 mark)**

(e) Nitrogen(V) oxide .... **(1 mark)**

(f) Chromium(VI) oxide .... **(1 mark)**

Redox reactions

1 Which of the following statements is correct?

☐ A Oxidation involves the loss of electrons and an increase in oxidation number.

☐ B Oxidation involves the gain of electrons and an increase in oxidation number.

☐ C Oxidation involves the gain of electrons and a decrease in oxidation number.

☐ D Oxidation involves the loss of electrons and a decrease in oxidation number. **(1 mark)**

Guided 2 Which of the following statements is correct?

☐ A In general, non-metals form negative ions with a decrease in oxidation number.

☐ B In general, metals form positive ions with a decrease in oxidation number.

☐ C In general, non-metals form negative ions with an increase in oxidation number.

☐ D̸ In general, metals form negative ions with a decrease in oxidation number.

> Option **D** cannot be correct because, in general, metals form positive ions.

(1 mark)

3 Acidified sodium dichromate(VI) is an oxidising agent capable of changing ethanol to ethanal.

(a) In terms of electrons, explain what an **oxidising agent** is.

.... **(1 mark)**

(b) During the reaction described above, Cr^{3+} ions are produced from $Cr_2O_7^{2-}$ ions. In terms of the change in oxidation number of chromium, explain whether this represents oxidation or reduction.

.... **(2 marks)**

4 The thermite reaction is useful for producing molten iron to weld railway track together:

$$2Al + Fe_2O_3 \rightarrow 2Fe + Al_2O_3$$

Explain why this reaction is a redox reaction.

.... **(2 marks)**

5 Calomel is mercury(I) chloride, a substance used in a certain type of electrode in chemistry. It decomposes when exposed to ultraviolet light:

$$Hg_2Cl_2 \rightarrow HgCl_2 + Hg$$

(a) Calculate the oxidation number(s) of mercury in each of the two products.

$HgCl_2$ ....   Hg .... **(2 marks)**

(b) Classify the reaction as redox or disproportionation, and explain your answer.

....

.... **(1 mark)**

6 Chlorine can displace bromine from bromide ions in solution:

$$Cl_2 + 2Br^- \rightarrow 2Cl^- + Br_2$$

> Are they oxidising agents or reducing agents, and why?

Explain the role of bromide ions in this reaction.

.... **(2 marks)**

Ionic half-equations

1 When magnesium powder is added to copper(II) sulfate solution, magnesium sulfate solution and copper metal are produced.

(a) Write a balanced half-equation for the oxidation of Mg to Mg^{2+} ions.

> You will need to add electrons on the right hand side of the equation.

.... **(1 mark)**

(b) Write a balanced half-equation for the reduction of Cu^{2+} ions to Cu.

> You will need to add electrons on the left hand side of the equation.

.... **(1 mark)**

(c) Use your answers to parts (a) and (b) to write a full ionic equation for the reaction.

.... **(1 mark)**

2 Use each of the following pairs of half-equations to write balanced full ionic equations.

(a) $Br_2 + 2e^- \rightarrow 2Br^-$ and $2I^- \rightarrow I_2 + 2e^-$

.... **(1 mark)**

(b) $NO_3^- + 2H^+ + e^- \rightarrow NO_2 + H_2O$

      and

$V^{3+} + H_2O \rightarrow VO^{2+} + 2H^+ + e^-$

> Check that you have cancelled H^+ and H_2O where you can.

.... **(2 marks)**

(c) $Cr_2O_7^{2-} + 14H^+ + 6e^- \rightarrow 2Cr^{3+} + 7H_2O$ and $H_2O_2 \rightarrow O_2 + 2H^+ + 2e^-$

.... **(2 marks)**

> **Guided**

3 Acidified potassium manganate(VII) solution is used to titrate solutions of iron(II) ions. During the reaction, manganate(VII) ions are reduced to manganese(II) ions, and iron(II) ions are oxidised to iron(III) ions.

(a) Complete and balance this half-equation.

$MnO_4^- +$ .... $H^+ +$ .... $e^- \rightarrow Mn^{2+} +$ .... H_2O **(1 mark)**

(b) The half-equation for the oxidation of Fe^{2+} ions is: $Fe^{2+} \rightarrow Fe^{3+} + e^-$
Use this, and your answer to part (a), to write the full equation for the reaction between manganate(VII) ions and iron(II) ions.

.... **(2 marks)**

4 A precipitate of silver chloride, AgCl, forms when silver nitrate solution, $AgNO_3$, is added to sodium chloride solution. Write the ionic equation for this reaction. Include state symbols.

.... **(2 marks)**

Reactions of Group 2 elements

1 Calcium reacts vigorously with water, producing effervescence and:

   ☐ **A** a red flame

   ☐ **B** a colourless solution

   ☐ **C** a cloudy white suspension

   ☐ **D** a pale green flame. **(1 mark)**

> **Guided**

2 (a) State the trend in first ionisation energy down Group 2.

   Going down Group 2, the first ionisation energy .... **(1 mark)**

   (b) Explain the trend in first ionisation energy down Group 2.

> You will need to consider factors such as atomic radius and shielding.

   ....

   .... **(2 marks)**

3 Barium reacts vigorously with water.

   (a) Write an equation for this reaction. Include state symbols in your answer.

   .... **(2 marks)**

   (b) Describe two observations that would be made in this reaction.

   ....

   .... **(2 marks)**

4 (a) Write an equation to describe the reaction between calcium and chlorine.

   .... **(1 mark)**

   (b) Explain, in terms of oxidation numbers, why this reaction is a redox reaction.

   ....

   .... **(2 marks)**

5 (a) Write an equation to describe the reaction between radium and oxygen.

   .... **(1 mark)**

   (b) Use your knowledge of the reactivity of the Group 2 elements to suggest **two** observations that might be made when radium is heated in oxygen.

   ....

   .... **(2 marks)**

Reactions of Group 2 compounds

1 Which of the following correctly describes trends down Group 2?

☐ **A** Hydroxide solubility increases and sulfate solubility decreases.

☐ **B** Hydroxide solubility decreases and sulfate solubility increases.

☐ **C** Hydroxide solubility decreases and sulfate solubility decreases.

☐ **D** Hydroxide solubility increases and sulfate solubility increases.   **(1 mark)**

2 Barium oxide reacts with water to produce barium hydroxide solution.

(a) Write an equation for this reaction.

....   **(1 mark)**

(b) Barium hydroxide solution can be used in titrations to determine the concentration of acids.

(i) Write an equation for the reaction between hydrochloric acid and barium hydroxide.

....   **(1 mark)**

(ii) Write an equation for the reaction between sulfuric acid and barium hydroxide.

....   **(1 mark)**

(c) One of the reactions described in part (b) will produce a precipitate.
Give the name and colour of this precipitate and write an ionic equation for its formation.

Name and colour ....

Ionic equation ....   **(3 marks)**

3 Milk of magnesia is a suspension of magnesium hydroxide in water, used as an antacid.

(a) Explain why magnesium hydroxide forms a suspension in water.

....   **(1 mark)**

(b) Hydrochloric acid is the acid in stomach acid.
Write an equation to show the reaction between Milk of Magnesia and stomach acid.

....   **(1 mark)**

4 Barium nitrate solution can be used to detect the presence of sulfate ions.
Explain why dilute nitric acid is added to the test solution before adding barium nitrate.

....   **(1 mark)**

5 Calcium oxide is an ingredient in lime mortar, a traditional mortar used to bind bricks together.
Suggest why it is wise to avoid skin contact with wet lime mortar.

....   **(1 mark)**

Stability of carbonates and nitrates

1   Explain what is meant by the term **thermal decomposition**.

.... **(1 mark)**

〉**Guided**〉 2   Which of the following substances are produced when lithium nitrate is heated?

☐ **A** lithium nitrite, nitrogen dioxide and oxygen

☐ **B** lithium nitrite and oxygen

☐ **C** lithium oxide, nitrogen dioxide and oxygen

☐ ~~**D**~~ lithium oxide and oxygen.

> Option **D** cannot be correct because it does not contain a nitrogen compound.

(1 mark)

3   Potassium nitrate, KNO_3, and calcium nitrate, $Ca(NO_3)_2$, decompose when heated.

(a) Write an equation for the decomposition of potassium nitrate.

.... **(2 marks)**

(b) Write an equation for the decomposition of calcium nitrate.

.... **(2 marks)**

4   Potassium carbonate and magnesium carbonate are both white solids.
What happens when they are heated strongly?

☐ **A** Potassium carbonate decomposes but magnesium carbonate does not decompose.

☐ **B** Magnesium carbonate decomposes but potassium carbonate does not decompose.

☐ **C** Neither carbonate decomposes.

☐ **D** Both carbonates decompose. **(1 mark)**

5   Limestone is mainly calcium carbonate. Cement is made by heating a mixture of powdered limestone and clay in a kiln.

(a) Write an equation for the decomposition of calcium carbonate.

.... **(1 mark)**

(b) Strontium carbonate is used in the manufacture of electroluminescent materials, such as those used in alarm clock faces. It must be heated first to decompose it to strontium oxide.

(i) Which substance, $CaCO_3$ or $SrCO_3$, needs the higher temperature to decompose?

.... **(1 mark)**

(ii) Explain the difference in stability between these two carbonates.

> Factors to consider include the size of the metal ion and how much it affects the anion.

....

....

.... **(3 marks)**

Had a go ☐ Nearly there ☐ Nailed it! ☐

Flame tests

1 (a) Describe how you would carry out a flame test on a solid compound.

....

....

....

....

Mention the apparatus and reagents you would need and describe what you would do with them.

(3 marks)

(b) Suggest a reason why this test would not detect magnesium chloride.

.... **(1 mark)**

2 Which of the following is the flame colour for barium compounds?

☐ **A** blue-green

☐ **B** blue

☐ **C** pale green

☐ **D** white **(1 mark)**

3 Complete the table below.

Group 1 cation	Flame colour
K^+	
	red-violet
Li^+	
	blue
Na^+	

(5 marks)

4 A student carried out a flame test on two white powders, strontium nitrate and calcium nitrate.

(a) State the flame test colours for strontium compounds and for calcium compounds.

Strontium ....

Calcium .... **(2 marks)**

(b) Suggest why it might be difficult to be certain which powder was which using a flame test.

.... **(1 mark)**

5 Group 1 chlorides produce coloured flames when:

☐ **A** the atoms lose outer electrons

☐ **B** electrons are excited to higher energy levels

☐ **C** electrons move from the chloride ions to the metal ions

☐ **D** excited electrons move from a higher to a lower energy level. **(1 mark)**

Properties of Group 7 elements

1   Which of these properties decreases going down Group 7?

☐ **A** Melting temperature

☐ **B** Electronegativity

☐ **C** Atomic number

☐ **D** Strength of intermolecular forces **(1 mark)**

Guided 2   (a) Describe the trend in boiling temperature down Group 7.

Going down Group 7, the boiling temperature .... **(1 mark)**

(b) Explain the trend in boiling temperature down Group 7.

....

....

> You will need to identify the bonds being broken during boiling, and the factors affecting their strength.

....

.... **(4 marks)**

3   Complete the table below.

Group 7 element	State at room temperature	Colour of vapour
chlorine		yellow-green
bromine		
iodine		

(5 marks)

4   The electronegativity of fluorine is different from the electronegativity of astatine.

(a) Which has the higher electronegativity, fluorine or astatine?

.... **(1 mark)**

(b) Explain why there is a difference in electronegativity between these two elements.

....

....

.... **(3 marks)**

5   If you try to draw bromine up into a plastic pipette, the warmth from your fingers can cause it to escape, even if you are not squeezing the pipette.
Suggest why this happens.

.... **(1 mark)**

Reactions of Group 7 elements

Guided 1 Going down Group 7 from chlorine to iodine:

☐ **A** the strength of the bond between atoms increases

☐ **B** the energy released when halide ions form decreases

☐ **C̶** the charge on the halide ion increases

☐ **D** the reactivity increases.

> Option **C** cannot be correct because all halide ions have a –1 charge.

(1 mark)

2 Caesium reacts spontaneously with chlorine, producing white smoke and a bright light.

(a) Name the product of this reaction, which causes the white smoke.

.... **(1 mark)**

(b) Write an equation to describe the reaction between caesium and chlorine.

.... **(1 mark)**

(c) Explain, in terms of oxidation numbers, why this reaction is a redox reaction.

....

.... **(2 marks)**

3 Halogens may react with halide ions in aqueous solution.

(a) Bromine and potassium iodide react in solution to produce potassium bromide and iodine.

(i) Write an equation for this reaction.

.... **(1 mark)**

(ii) Write an ionic equation for the reaction.

.... **(1 mark)**

(iii) What would you see if some cyclohexane was added to the mixture, then shaken?

....

.... **(2 marks)**

(iv) What difference would you see in part (iii) if potassium chloride was used instead of potassium bromide?

.... **(1 mark)**

(b) Suggest what you would observe when fluorine gas is blown briefly onto a piece of filter paper, dampened with potassium iodide solution. Give reasons for your answer.

....

.... **(2 marks)**

Reactions of chlorine

1   Explain what is meant by the term **disproportionation**.

.... **(1 mark)**

2   Bromine dissolves in water and reacts with it:

$$Br_2(aq) + H_2O(l) \rightleftharpoons HBr(aq) + HBrO(aq)$$

Which statement about this reaction is **not** correct?

☐ **A** The oxidation number of bromine in Br_2 is 0.

☐ **B** The oxidation number of some bromine atoms increases.

☐ **C** The oxidation number of some bromine atoms decreases.

☐ **D** The mixture is alkaline.   **(1 mark)**

3   Explain why chlorine is used in water treatment.

.... **(1 mark)**

4   Chlorine reacts with cold, dilute sodium hydroxide solution:

$$Cl_2(aq) + 2NaOH(aq) \rightarrow NaCl(aq) + NaClO(aq) + H_2O(l)$$

(a) Write an ionic equation for the reaction. State symbols are not needed.

....

....

> Sodium ions are spectator ions in the reaction.

(1 mark)

(b) (i) State the name of the product, NaClO.

....

....

> You will need to work out the oxidation number of chlorine in this compound.

(1 mark)

(ii) State a use for NaClO.

.... **(1 mark)**

5   Bromine reacts with hot concentrated sodium hydroxide solution to form sodium bromate(V), sodium bromide and water. The reaction is similar to the one between chlorine and hot concentrated sodium hydroxide solution.

(a) The formula for sodium bromide is NaBr. Give the formula for sodium bromate(V).

.... **(1 mark)**

(b) Write an equation for the reaction described above. State symbols are not needed.

.... **(2 marks)**

(c) State the type of reaction and explain your answer in terms of oxidation numbers.

....

....

.... **(3 marks)**

Halides as reducing agents

1 Explain what is meant by the term **reducing agent**.

.... **(1 mark)**

2 Iodine is produced when concentrated sulfuric acid is added to solid sodium iodide, but no chlorine is produced when the same acid is added to solid sodium chloride. This is because:

☐ **A** chloride ions are weaker reducing agents than iodide ions

☐ **B** sulfuric acid is a strong acid

☐ **C** hydriodic acid is a weak acid

☐ **D** chlorine is more volatile than iodine.   **(1 mark)**

3 Hydrogen bromide is formed when concentrated sulfuric acid reacts with sodium bromide:

$$NaBr + H_2SO_4 \rightarrow HBr + NaHSO_4$$

(a) Explain, in terms of oxidation numbers, why this is not a redox reaction.

.... **(1 mark)**

(b) Describe what is observed when hydrogen bromide is released in the reaction.

.... **(1 mark)**

(c) Bromine and sulfur dioxide form when hydrogen bromide reacts with concentrated sulfuric acid.

(i) Write an equation for this reaction. | Water is also formed. |

.... **(1 mark)**

(ii) State the role of the bromide ions in this reaction.

.... **(1 mark)**

4 Iodine and sulfur form when hydrogen iodide reacts with concentrated sulfuric acid:

$$6HI + H_2SO_4 \rightarrow S + 3I_2 + 4H_2O$$

(a) Write the half-equations for:

(i) the oxidation of iodide ions

.... **(1 mark)**

(ii) the reduction of sulfuric acid.

| You will need hydrogen ions on the left. |

.... **(1 mark)**

(b) State the role of the concentrated sulfuric acid in this reaction.

.... **(1 mark)**

(c) Name the product formed in the reaction between sodium iodide and concentrated sulfuric acid that has a smell of rotten eggs.

.... **(1 mark)**

Other reactions of halides

> **Guided**

1 Which statement about hydrogen halides is correct?

☐ **A** They exist as coloured gases.

☐ **B** They dissolve in water to form acidic solutions.

☐ **C̶** They react with barium chloride solution to form precipitates.

☐ **D** Hydrofluoric acid is a strong acid.

> Option **C** cannot be correct because this describes the test for sulfate ions, not halide ions.

(1 mark)

2 Solutions containing sodium fluoride or sodium chloride can be distinguished by observing the results of a simple test. Dilute nitric acid is added, then silver nitrate solution.

(a) Explain why dilute nitric acid is added.

.... **(1 mark)**

> **Guided**

(b) Explain how you would use the results of the test to distinguish between NaF and NaCl.

With NaF I would observe ....

but with NaCl I would observe .... **(2 marks)**

3 Potassium bromide solution reacts with silver nitrate solution to form a precipitate.

(a) State the colour of the precipitate.

.... **(1 mark)**

(b) Write an equation for the reaction, including state symbols in your answer.

.... **(1 mark)**

(c) Explain how you could use aqueous ammonia to confirm the presence of bromide ions.

....

.... **(2 marks)**

4 Silver chloride and silver iodide are both solids at room temperature.

(a) State the colours of these two compounds.

Silver chloride ....

Silver iodide .... **(2 marks)**

(b) Explain how you could use the aqueous ammonia to distinguish between the two solids.

....

....

....

> Explain what you would do, and what you would see for each silver halide.

(3 marks)

31

Exam skills 3

1 (a) A solution of bromine is added to potassium iodide solution.

  (i) Write an ionic equation for the reaction.

  ....

  (ii) Explain, in terms of electrons, why this is a redox reaction.

  ....

  ....

  (iii) Describe what you would expect to see if the reaction mixture is added to cyclohexane.

  .... **(7 marks)**

 (b) A mixture of $NaCl$ and $NaClO_3$ forms when $NaClO$ is heated above $75\,°C$.

  (i) Write a balanced equation for the reaction.

  ....

  (ii) Explain, in terms of oxidation numbers, why this is a disproportionation reaction.

  ....

  .... **(4 marks)**

 (c) Sulfur is formed when concentrated sulfuric acid is added to potassium iodide.

  (i) Write the half-equation for the formation of sulfur from concentrated sulfuric acid.

  ....

  (ii) In a different redox reaction between concentrated sulfuric acid and potassium iodide, a black solid and a toxic gas with an unpleasant smell form. Identify these products, and write an equation for the reaction.

  Name of black solid: ....

  Equation: .... **(6 marks)**

 (d) Describe how you could test a solution for the presence of iodide ions.

  ....

  ....

  .... **(4 marks)**

Moles and molar mass

> **Guided**

1 Define the term **mole**.

One mole is the amount of substance that contains ....

.... **(3 marks)**

2 Assuming that the Avogadro constant is 6.0×10^{23} mol^{-1}, the number of atoms in 1 mol of ethene, C_2H_4, is:

☐ **A** 1.0×10^{23}

☐ **B** 6.0×10^{23}

☐ **C** 3.6×10^{23}

☐ **D** 3.6×10^{24} **(1 mark)**

3 State the units used (if any) for:

(a) Amount of substance .... **(1 mark)**

(b) Molar mass .... **(1 mark)**

(c) Relative formula mass .... **(1 mark)**

4 Use these relative atomic masses, A_r, to help you answer the following questions.

Element	H	C	N	O	Na	Cl
A_r	1.0	12.0	14.0	16.0	23.0	35.5

Calculate the mass of the following substances.

(a) 1.0 mol of carbon dioxide, CO_2.

> mass = amount × molar mass

.... **(2 marks)**

(b) 0.2 mol of sodium chloride, NaCl.

.... **(2 marks)**

(c) 2.5 mol of dinitrogen tetroxide, N_2O_4.

.... **(2 marks)**

(d) 5.0×10^{-3} mol of methanol, CH_3OH.

.... **(2 marks)**

5 Calculate the mass of hydrogen atoms in 4.0 mol of water, H_2O.

....

.... **(1 mark)**

Empirical and molecular formulae

1 Define the term **empirical formula**.

....

.... **(2 marks)**

▷**Guided**▷ **2** A blue-violet solid contains 0.874 g of chromium, 0.706 g of nitrogen and 2.42 g of oxygen. Calculate its empirical formula.

Symbol	Cr	N	O
Mass /g			
Molar mass			
g ÷ molar mass			
Divide by smallest value			

Empirical formula is .... **(3 marks)**

3 0.365 g of magnesium combined completely with nitrogen to produce 0.505 g of a white solid. Calculate the empirical formula of this product.

....

....

....

.... **(3 marks)**

4 A certain hydrocarbon contains 85.7% carbon by mass. Calculate:

> Hydrocarbons are compounds of hydrogen and carbon only.

(a) The percentage of hydrogen .... **(1 mark)**

(b) The empirical formula of the hydrocarbon.

....

....

.... **(2 marks)**

(c) The molar mass of the hydrocarbon is 42.0 g mol^{-1}. Determine its molecular formula.

....

.... **(2 marks)**

Reacting masses calculations

Guided

1 Calculate the mass of carbon dioxide produced from the combustion of 3.0 g of carbon:

$$C(s) + O_2(g) \rightarrow CO_2(g)$$

molar mass of C = ... molar mass of CO_2 = ...

amount of C = mass of C ÷ molar mass of C = ...

amount of CO_2 produced in mol = ...

mass of CO_2 = amount of CO_2 × molar mass of CO_2 = ... **(3 marks)**

Maths skills

2 Calculate the mass of oxygen needed to react completely with 1.50 g of magnesium:

$$2Mg(s) + O_2(g) \rightarrow 2MgO(s)$$

Give your answer to 3 significant figures.

> Look at the fourth significant figure. Round up if this is 5 or more. For example, 4.321 is 4.32 to 3 significant figures, but 4.325 is 4.33 to 3 significant figures.

....

....

.... **(3 marks)**

3 One of the reactions that happens in a blast furnace involves iron(III) oxide and carbon monoxide:

$$Fe_2O_3(s) + 3CO(g) \rightarrow 2Fe(l) + 3CO_2(g)$$

> 1 tonne = 1×10^6 g

Calculate the mass of iron that can be produced from 3.00 tonnes of carbon monoxide and an excess of iron(III) oxide.

....

....

.... **(3 marks)**

4 When an iron oxide is heated with excess carbon, 2.79 g of iron and 1.10 g of carbon dioxide form.

(a) Calculate the amount (in mol) of each product.

Amount of iron = ...

Amount of carbon dioxide = ... **(1 mark)**

(b) Show that the equation for the reaction could be $2FeO + C \rightarrow 2Fe + CO_2$.

....

.... **(1 mark)**

Gas volume calculations

1 At the same temperature and pressure, which of the following gases occupies the greatest volume?

(Use these relative atomic masses: C = 12.0, N = 14.0, O = 16.0, Ne = 20.2)

☐ **A** 10 g of neon

☐ **B** 10 g of carbon monoxide

☐ **C** 10 g of nitrogen

☐ **D** 10 g of oxygen

> 1 mole of any gas occupies 24.0 dm³ at room temperature and pressure.

(1 mark)

2 Hydrogen and chlorine react together to produce hydrogen chloride:

$$H_2(g) + Cl_2(g) \rightarrow 2HCl(g)$$

(a) Calculate the volume of hydrogen chloride that could be produced from 50 cm³ of hydrogen and excess chlorine. Assume that the temperature and pressure do not change.

.... **(1 mark)**

(b) Calculate the total gas volume from a complete reaction involving 100 cm³ of hydrogen and 150 cm³ of chlorine. Assume that the temperature and pressure do not change.

....

.... **(2 marks)**

3 A camping gas cylinder holds 190 g of butane, which burns completely in oxygen to form carbon dioxide and water:

$$C_4H_{10}(g) + 6\tfrac{1}{2}O_2(g) \rightarrow 4CO_2(g) + 5H_2O(l)$$

Assume that the molar volume of a gas, V_m, is 24 dm³.
Give your answers to 3 significant figures.

(a) Calculate the amount (in mol) of butane in the gas cylinder.

....

.... **(2 marks)**

(b) Calculate the amount (in mol) of oxygen needed to react completely with the butane.

.... **(1 mark)**

(c) Use your answer to part (b) to calculate the volume of oxygen needed.

.... **(1 mark)**

(d) Calculate the maximum volume of carbon dioxide produced.

.... **(1 mark)**

(e) Calculate the maximum mass of water produced.

....

.... **(2 marks)**

Concentrations of solutions

1 Which expression could you use to calculate mass concentration in $g\,dm^{-3}$ from molar concentration in $mol\,dm^{-3}$?

    ☐ **A** molar concentration ÷ molar mass

    ☐ **B** molar concentration × molar mass

    ☐ **C** molar concentration + molar mass

    ☐ **D** molar concentration − molar mass   **(1 mark)**

2 (a) Convert these volumes to dm^3.

> There are $1000\,cm^3$ in $1\,dm^3$.

    (i) $2500\,cm^3$ .... **(1 mark)**

    (ii) $50\,cm^3$ .... **(1 mark)**

  (b) Convert these volumes to cm^3.

    (i) $1.25\,dm^3$ .... **(1 mark)**

    (ii) $0.02\,dm^3$ .... **(1 mark)**

3 Calculate the molar concentrations (in $mol\,dm^{-3}$) of the following solutions.

  (a) $2\,mol$ of solute in $4\,dm^3$ of solution.

  .... **(1 mark)**

  (b) $0.25\,mol$ of solute in $100\,cm^3$ of solution.

  .... **(1 mark)**

4 Calculate the amount of solute dissolved in the following solutions.

  (a) $0.5\,dm^3$ of a $2\,mol\,dm^{-3}$ solution.

  .... **(1 mark)**

  (b) $200\,cm^3$ of a $0.25\,mol\,dm^{-3}$ solution.

  .... **(1 mark)**

5 Anhydrous sodium carbonate, Na_2CO_3, was used to make a $0.1\,mol\,dm^{-3}$ standard solution.

  (a) Calculate the amount of sodium carbonate needed to make $250\,cm^3$ of this solution.

  .... **(1 mark)**

  (b) Calculate the molar mass of sodium carbonate.

  .... **(1 mark)**

  (c) Use your answers to parts (a) and (b) to calculate the mass of sodium carbonate needed.

  .... **(1 mark)**

Doing a titration

1   A student carries out a titration in which dilute hydrochloric acid is added from a burette to sodium hydroxide solution in a conical flask. Phenolphthalein indicator is used. This is a weak acid that is pink in alkaline solution and colourless in acidic solution.

(a)   Explain why the student should use a white tile.

So that the colour change of the ✓ **(1 mark)**
indicator is obvious.

(b)   The student adds a full dropper of phenolphthalein to the conical flask, rather than just a few drops. Suggest the effect, if any, on the titre obtained. Explain your answer.

The titre will be inaccurate as the solution ⚡
would be pink before it is neutralised. **(2 marks)**

(c)   The student forgets to make sure that the burette is vertical and it is tilted so that the top is away from them. Suggest the effect, if any, on the readings. Explain your answer.

The student will not be able to read ⚡
the burette accurately and the titre will be **(1 mark)**
inaccurate.

(d)   After adding acid to the burette, the student forgets to open the tap briefly, so air is left in the tip of the burette. Suggest the effect, if any, on the titre obtained. Explain your answer.

The burette reading will be inaccurate as
it will include an air bubble and so the **(1 mark)**
actual amount of acid in the burette will be
less than the amount measured

2   Explain what is meant is by the following terms.

(a)   End point the amount of substance in the
burette when the substance is the conical **(1 mark)**
flask has been neutralised

(b)   Concordant results titres that are within
0.1 cm³ of each other. ✓ **(1 mark)**

3   The error when using a 25 cm³ pipette is 0.06 cm³. Calculate the percentage error when using the pipette.

0.24 %. ✓ **(1 mark)**

4   The error in a single burette reading is 0.05 cm³. When you calculate a titre, you use two readings, so the error in the titre is 0.10 cm³. Calculate the percentage error for a titre of 22.95 cm³ and give your answer to 2 significant figures.

0.44 %. ✓ **(1 mark)**

Titration calculations

1 The carbonate of a Group 1 metal reacts with hydrochloric acid:

$$M_2CO_3 + 2HCl \rightarrow 2MCl + H_2O + CO_2$$

1.55 g of M_2CO_3 was dissolved in water to make 250 cm³ of solution. In a titration, 29.20 cm³ of 0.100 mol dm⁻³ hydrochloric acid reacted completely with 25.0 cm³ of this solution.
In the following questions, give your answers to 3 significant figures.

(a) Calculate the amount (in mol) of HCl in the mean titre.

.... **(1 mark)**

(b) Calculate the amount (in mol) of M_2CO_3 that reacts with the amount calculated in part (a).

.... **(1 mark)**

(c) Use your answer to part (b) to calculate the amount of M_2CO_3 in the original 1.55 g.

> Remember that the 1.55 g of M_2CO_3 was in a 250 cm³ solution.

.... **(1 mark)**

(d) Use your answer to part (c) and the mass of 1.55 g used to calculate the molar mass of M_2CO_3.

> You know the amount in mol of M_2CO_3 and its mass.

.... **(1 mark)**

(e) Use your answer to part (d) to calculate the molar mass of M, and to name the carbonate.

> You will need to calculate the molar mass of CO_3 as part of your answer.

....

.... **(2 marks)**

2 Nitric acid reacts with sodium hydroxide:

$$HNO_3 + NaOH \rightarrow NaNO_3 + H_2O$$

What volume of 0.50 mol dm⁻³ nitric acid is needed to neutralise 0.010 mol of sodium hydroxide?

☐ **A** 5 cm³

☐ **B** 20 cm³

☐ **C** 25 cm³

☐ **D** 50 cm³ **(1 mark)**

3 Ethanedioic acid reacts with sodium hydroxide:

$$H_2C_2O_4 + 2NaOH \rightarrow Na_2C_2O_4 + 2H_2O$$

In a titration, 27.50 cm³ of 0.0500 mol dm⁻³ ethanedioic acid reacted completely with 25.0 cm³ of a sodium hydroxide solution. Calculate the concentration of the sodium hydroxide.

....

....

.... **(3 marks)**

Atom economy

Guided 1 During the manufacture of calcium oxide from limestone, the theoretical yield was 2.0 tonnes. Calculate the percentage yield if the actual yield was 1.6 tonnes.

$$\text{percentage yield} = \frac{\text{actual yield}}{\text{theorectical yield}} \times 100 = \text{...}$$ **(1 mark)**

2 Copper(II) sulfate can be prepared from copper(II) oxide and sulfuric acid:

$$CuO + H_2SO_4 \rightarrow CuSO_4 + H_2O$$

(a) Calculate the molar masses of CuO, $CuSO_4$ and H_2O in $g\,mol^{-1}$.

CuO ....

CuSO₄ ....

> Use the Periodic Table to obtain the relative atomic masses of each element.

H₂O ....   **(1 mark)**

(b) Calculate the theoretical yield of copper(II) sulfate from 2.5 g of copper(II) oxide.

....

.... **(1 mark)**

(c) The actual yield was 3.5 g of copper(II) sulfate. Calculate the percentage yield.

.... **(1 mark)**

Guided 3 Methanol (molar mass $32.0\,g\,mol^{-1}$) can be manufactured by reacting carbon monoxide (molar mass $28.0\,g\,mol^{-1}$) with hydrogen:

$$CO + 2H_2 \rightarrow CH_3OH$$

What is the percentage yield if 28 tonnes of carbon monoxide produces 20 tonnes of methanol?

☐ ~~A~~ 140%

☐ **B** 100%

> Option **A** cannot be correct because the yield cannot be more than 100%

☐ **C** 71%

☐ **D** 62.5%   **(1 mark)**

4 Hydrogen can be manufactured by the electrolysis of water:

$$2H_2O \rightarrow 2H_2 + O_2$$

(a) Calculate the atom economy for making hydrogen this way.

....

.... **(2 marks)**

(b) Suggest one way to improve the atom economy of this process.

.... **(1 mark)**

5 Which one of the four reactions below has the highest atom economy for making oxygen?

☐ **A** $BaO_2 \rightarrow Ba + O_2$

☐ **B** $2HgO \rightarrow 2Hg + O_2$

☐ **C** $2H_2O \rightarrow 2H_2 + O_2$

☐ **D** $2H_2O_2 \rightarrow 2H_2O + O_2$ **(1 mark)**

Exam skills 4

1   Baking soda contains sodium hydrogencarbonate, $NaHCO_3$. A 1.48 g sample of baking
    soda was dissolved in 75.0 cm^3 of 0.500 $mol\,dm^{-3}$ hydrochloric acid. The mixture was
    added to a volumetric flask and made up to 250 cm^3 with deionised water. The mean titre
    was 25.40 cm^3 when 25.0 cm^3 portions of this mixture were titrated with 0.100 $mol\,dm^{-3}$
    sodium hydroxide solution.
    The equations for the reactions involved are:

$$NaHCO_3(s) + HCl(aq) \rightarrow NaCl(aq) + H_2O(l) + CO_2(g)$$

$$HCl(aq) + NaOH(aq) \rightarrow NaCl(aq) + H_2O(l)$$

(a)   Calculate the number of moles of hydrochloric acid that reacted with the
      sodium hydroxide solution.

.... **(1 mark)**

(b)   Calculate the number of moles of hydrochloric acid that did **not** react with the
      sodium hydrogencarbonate in the baking soda.

.... **(1 mark)**

(c)   Calculate the number of moles of hydrochloric acid added to the baking soda.

.... **(1 mark)**

(d)   Use your answers to parts (b) and (c) to calculate the number of moles of
      sodium hydrogencarbonate that reacted with the hydrochloric acid.

....

....

.... **(2 marks)**

(e)   Calculate the percentage of sodium hydrogencarbonate in the baking soda.
      Give your answer to three significant figures.

....

....

.... **(3 marks)**

(f)   Calculate the volume of carbon dioxide produced when the hydrochloric acid
      was added to the baking soda. Give your answer to three significant figures.
      (Molar volume of gas = 24 $dm^3\,mol^{-1}$)

....

.... **(2 marks)**

Had a go ☐ Nearly there ☐ Nailed it! ☐

Alkanes

1 Which of these is the general formula for alkanes?

☐ **A** C_nH_n

☐ **B** C_nH_{2n}

☐ **C** C_nH_{2n-2}

☒ **D** C_nH_{2n+2} **(1 mark)**

2 Explain why ethane may be described as a saturated hydrocarbon.

Because it has as many hydrogens as possible, it doesn't contain a double bond. **(2 marks)**

* Because it is a compound made up of only hydrogen and carbon, and it doesn't contain a C=C double bond.

3 Butane may be represented by the structural formula, $CH_3CH_2CH_2CH_3$.

(a) Write the molecular formula for butane.

C_4H_{10} **(1 mark)**

(b) Write the empirical formula for butane.

C_2H_5 **(1 mark)**

(c) Draw the displayed formula and skeletal formula for butane in the boxes below.

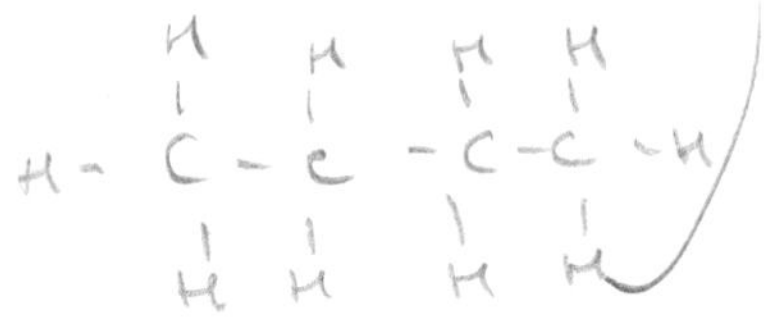

displayed formula

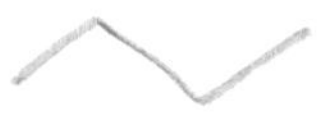

skeletal formula

(2 marks)

4 Dodecane is an unbranched alkane with 12 carbon atoms.

(a) Write the molecular formula for docdecane.

$C_{12}H_{26}$ **(1 mark)**

(b) Write the structural formula for docdecane.

$CH_3(CH_2)_{10}CH_3$ **(1 mark)**

Multiple CH_2 groups may be put inside brackets.

* 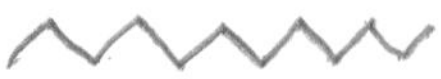

5 The diagram shows the skeletal formula for an alkane.

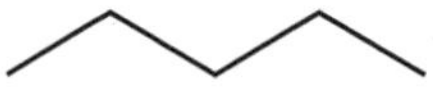

Name this alkane and write its molecular formula.

propane *pentane **(2 marks)**

Isomers of alkanes

1 Isomers have different:

☐ **A** molar masses

☒ **B** molecular formulae

☐ **C** skeletal formulae

☐ **D** empirical formulae.   **(1 mark)**

2 Name the alkane represented by this displayed formula and write its structural formula.

Name ....2,2 di methyl butane......

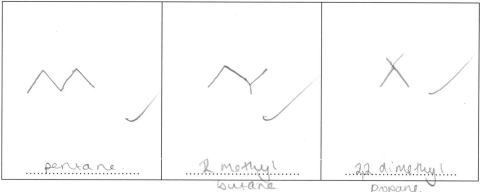

Structural formula ....✗............................. **(2 marks)**

3 There are three isomers with the molecular formula C_5H_{12}.

(a) In the boxes below, draw the skeletal formula of each isomer and write their names.

pentane.	2 methyl butane	2,2 dimethyl propane.

(6 marks)

(b) Name the type of isomerism shown by these molecules.

.........Structural isomerism.......... **(1 mark)**

4 How many isomers are there with the molecular formula C_6H_{12}? $C-C-C-C-C-C$

☐ **A** 4

☒ **B** 5

☐ **C** 6

☐ **D** 7

> You may need
> to draw them
> out first.

(1 mark)

5 Name the isomer of C_7H_{16} that has:

(a) three side chains ....1,2,3 tri methyl butane..... **(1 mark)**

(b) one side chain. .....methyl hexane..... **(1 mark)**

Alkenes

1   Which of these is the general formula for alkenes?

   ☐ **A** C_nH_n

   ☐ **B** C_nH_{2n}

   ☐ **C** C_nH_{2n-2}

   ☒ **D** C_nH_{2n+2} **(1 mark)**

2   Hex-1-ene is an unsaturated hydrocarbon. It is a liquid at room temperature.

   (a) Explain why it is described as unsaturated.

   Because it has a C=C bond and so has
   less than the maximum number of hydrogen atoms.   **(1 mark)**

   (b) Describe a chemical test you would carry out to distinguish between hexane and hex-1-ene.

   ....

   ....

   > State the reagent you would use, and the observations for each hydrocarbon.

                                                                   (2 marks)

3   (a) Write a structural formula for propene.

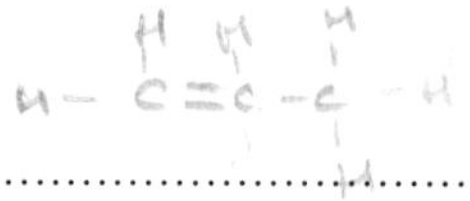

   ....   **(1 mark)**

   (b) Draw the displayed formula and skeletal formula for ethene in the boxes below.

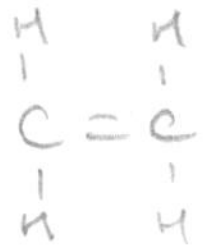 displayed formula	skeletal formula

                                                                   (2 marks)

4   There are two isomers with the molecular formula C_3H_6. One is saturated and the other is unsaturated. In the boxes below, draw their displayed formulae and skeletal formulae, and write their names.

	Saturated hydrocarbon	**Unsaturated hydrocarbon**
Displayed formula	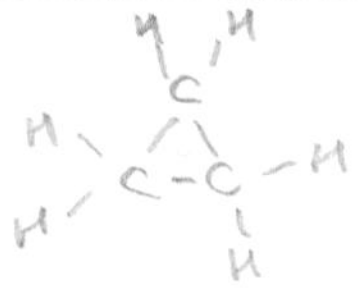	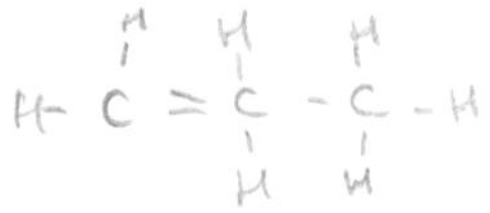
Skeletal formula	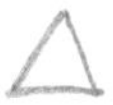	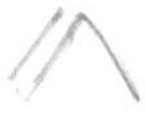
Name	cyclopropane.	propene

                                                                   (6 marks)

Isomers of alkenes

1 What is the name of this alkene?

$$H_3C—C=CH—CH_3$$
$$\quad\quad |$$
$$\quad\quad CH_3$$

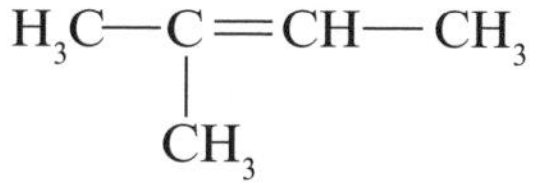

☐ **A** 2-methylbut-3-ene

☐ **B** 3-methylbut-2-ene

☐ **C** 3-methylbut-3-ene

☒ **D** 2-methylbut-2-ene   **(1 mark)**

2 There are two stereoisomers of C_4H_8.

(a) Explain why stereoisomerism can occur in alkenes.

...Because the alkyl groups can occur
on either side of the C=C bond in
an alkane.   **(2 marks)**

(b) (i) In the boxes below, complete the displayed formula for *E*-but-2-ene and for *Z*-but-2-ene.

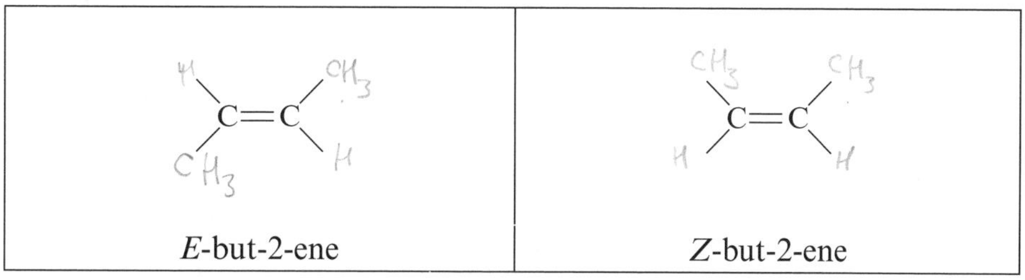

| *E*-but-2-ene | *Z*-but-2-ene |

(2 marks)

(ii) Name the two isomers using the *cis–trans* naming system.

E-but-2-ene ...trans but 2 ene...   *Z*-but-2-ene ...cis but-2-ene...   **(1 mark)**

(c) But-1-ene and its chain isomer do not have stereoisomers.

(i) Name the chain isomer.

...2 methyl prop-1-ene.   **(1 mark)**

(ii) Explain why these two isomers do not have stereoisomers.

...Because the C=C bond occurs on
the first carbon atom, so the alkyl groups
do not occur on either side of it.   **(1 mark)**

3 What is the name of this alkene?

$$H_3C\quad\quad CH_2CH_3$$
$$\quad\searrow\!C=C\!\nearrow$$
$$H\quad\quad CH_2CH_2CH_3$$

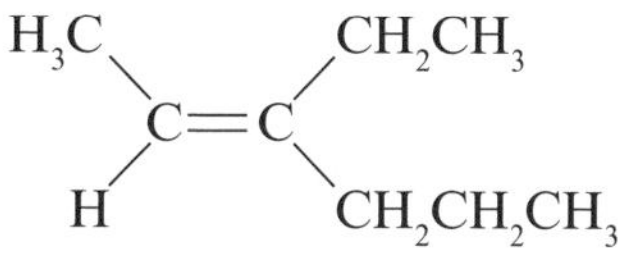

☐ **A** *Z*-3-propylpent-2-ene

☐ **B** *E*-4-ethylhex-4-ene

☒ **C** *E*-3-ethylhex-2-ene

☐ **D** *Z*-3-ethylhex-2-ene   **(1 mark)**

Using crude oil

1 Crude oil is separated into fractions by fractional distillation.

(a) Which property of hydrocarbon molecules allows crude oil to be separated into fractions?

...... Different boiling points ...... **(1 mark)**

A mixture of compounds with similar boiling temperature.

(b) In fractional distillation, what is meant by the term 'fraction'?

...... A 'fraction' is the part of the crude oil made up of one type of molecule. ...... **(1 mark)**

2 Which of the following crude oil fractions has the highest boiling temperature?

☐ **A** refinery gas

☒ **B** bitumen ✓

☐ **C** diesel oil

☐ **D** petrol **(1 mark)**

3 Ethene, C_2H_4, is obtained by cracking of crude oil fractions.

(a) Write an equation to show the cracking of hexane, C_6H_{14}, to produce ethene and one other hydrocarbon.

...... $C_6H_{14} \rightarrow C_2H_4 + C_4H_{10}$ . ... **(1 mark)**

(b) Give one economic reason why cracking is carried out.

...... The smaller hydrocarbons are more desirable because it is more useful ie, as fuel ...... **(1 mark)**

4 The hydrocarbon shown on the right is a product of reforming.

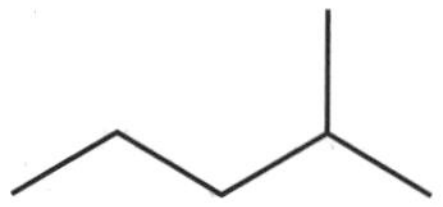

(a) Name this hydrocarbon.

...... 2 methyl pentane. ...... **(1 mark)**

(b) The hydrocarbon in part (a) was produced from an unbranched hydrocarbon. Name this hydrocarbon and state its molecular formula.

...... hexane C_6H_{14}. ...... **(2 marks)**

(c) The reforming of the hydrocarbon in part (b) can also produce an unbranched cyclic alkane with the same number of carbon atoms.

(i) Name this hydrocarbon and state its molecular formula.

...... Cyclohexane C_6H_{12} ...... **(2 marks)**

(ii) Identify the other product made in this reforming reaction.

...... H_2 hydrogen gas. ...... **(1 mark)**

✳ Hydrocarbons as fuels

1 The composition of natural gas is mainly methane, CH_4.

(a) Write an equation for the complete combustion of methane.

$$CH_4 + 2O_2 \rightarrow CO_2 + 2H_2O$$

.... **(1 mark)**

(b) Explain the main environmental problem caused by a product of complete combustion of methane and other alkanes.

It produces carbon dioxide which is
a greenhouse gas.

.... **(2 marks)**

2 Hexane, C_6H_{14}, typically forms about 3% of petrol.

(a) Explain why incomplete combustion of hexane can happen.

If not enough oxygen present

.... **(1 mark)**

(b) Write an equation for the incomplete combustion of hexane to form water and equal amounts of carbon monoxide and a solid product.

$$C_6H_{14} + 5O_2 \rightarrow 3CO + 7H_2O + 3C$$

.... **(2 marks)**

(c) State two problems arising from incomplete combustion of hexane and other alkanes.

It can produce carbon monoxide which
is toxic. carbon can cause breathing
problems

.... **(2 marks)**

3 Which of the following pollutants is produced as a result of impurities in fuels?

☒ **A** sulfur dioxide

☐ **B** unburned hydrocarbons

☐ **C** carbon particulates

☐ **D** NO_x

(1 mark)

4 Oxides of sulfur and nitrogen are formed during the combustion of alkane fuels.
State and explain the environmental problem caused by these gases.

Acid rain, these oxides have acidic
properties and can dissolve in clouds. When
it rains this can cause damage to
buildings → q

.... **(3 marks)**

5 (a) Write an equation to show how carbon monoxide and nitrogen monoxide may be converted into less harmful products in a catalytic converter.

$$2CO + 2NO \rightarrow 2CO_2 + N_2$$

.... **(1 mark)**

(b) State the type of reaction involved .... Redox. .... **(1 mark)**

Alternative fuels

1 Which of the following is **not** an alternative fuel?

☐ **A** biodiesel

☐ **B** bioalcohol

☐ **C** natural gas

☐ **D** wood chippings **(1 mark)**

⟩Guided⟩ **2** One of the advantages of alternative fuels compared to fossil fuels is that they are closer to being carbon neutral. What is meant by the term carbon neutral?

An activity that has no .... to the atmosphere. **(1 mark)**

3 Alternative fuels are renewable whereas fossil fuels are described as non-renewable.

(a) Explain what is meant by the term **renewable**.

.... **(1 mark)**

(b) Explain what is meant by the term **non-renewable**.

....

| You should not just write 'the opposite of renewable' or something similar. |

.... **(1 mark)**

4 Ethanol can be mixed with petrol for use as a vehicle fuel. There are two main ways to make it:

Process 1 $C_6H_{12}O_6 \rightarrow 2CH_3CH_2OH + 2CO_2$

Process 2 $CH_2{=}CH_2 + H_2O \rightarrow CH_3CH_2OH$

(a) Different raw materials are needed for each process. For each process, name one suitable raw material.

Process 1 ....   Process 2 .... **(2 marks)**

(b) Suggest **two** advantages of Process 1 over Process 2.

....

.... **(2 marks)**

(c) Suggest **two** disadvantages of Process 1 over Process 2.

....

.... **(2 marks)**

5 Biodiesel is manufactured using a strong base, an alcohol and vegetable oils. State **one** source of suitable vegetable oils.

.... **(1 mark)**

Alcohols and halogenoalkanes

1 The chloroalkanes form a homologous series. State three features of a homologous series.

1 ....

2 ....

3 .... **(3 marks)**

2 (a) Name the alcohols represented by these formulae. **(3 marks)**

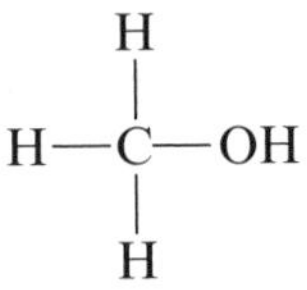

(i) .... (ii) .... (iii) ....

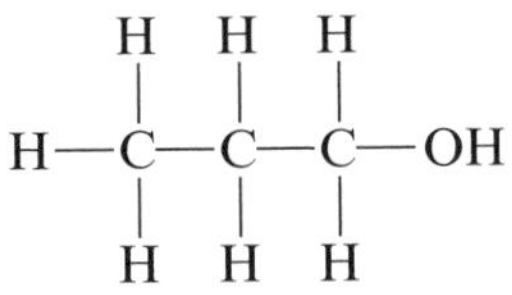

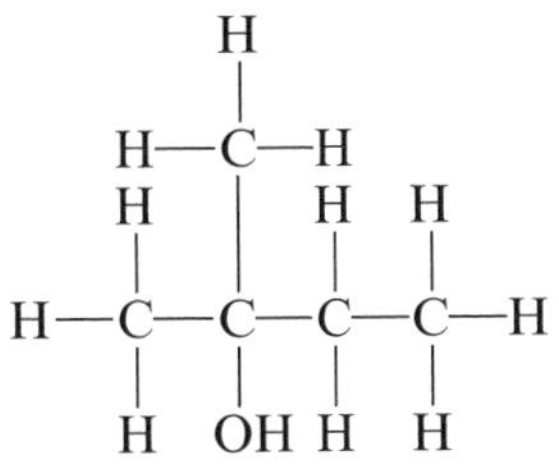

(b) Write molecular formulae and structural formulae for compounds (ii) and (iii) above.

> The molecular formula for (i) is CH_4O and its structural formula is CH_3OH.

Molecular formula (ii) ....

Structural formula (ii) .... **(2 marks)**

Molecular formula (iii) ....

Structural formula (iii) .... **(2 marks)**

(c) Name the functional group, −OH, present in alcohols.

.... **(1 mark)**

3 What is the systematic name for the compound represented by this skeletal formula?

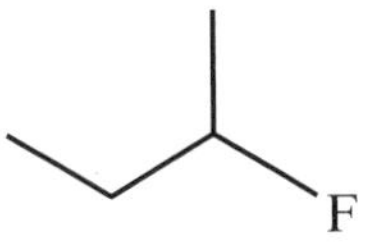

☐ **A** 1-fluoro-1-methylpropane

☐ **B** 2-fluorobutane

☐ **C** 3-fluorobutane

☐ **D** 1-fluoro-2-methylbutane **(1 mark)**

4 In the boxes below, draw skeletal formulae to represent each named halogenoalkane.

(a) 1-chloropropane	(b) 2-bromo-2-methylpropane	(c) 2-iodo-3-methylbutane

(3 marks)

Substitution reactions of alkanes

1   What type of species is formed by homolytic bond fission?

☐ **A** radical

☐ **B** ion

☐ **C** electrophile

☐ **D** nucleophile   **(1 mark)**

2   Methane and bromine react together to produce bromomethane and hydrogen bromide:

$$CH_4 + Br_2 \rightarrow CH_3Br + HBr$$

The mechanism is similar to the one for the reaction between methane and chlorine.

(a) Name the mechanism involved.

.... **(1 mark)**

(b) Outline the mechanism for the reaction between methane and bromine.

(i)   Initiation step .... **(1 mark)**

(ii) First propagation step .... **(1 mark)**

(iii) Second propagation step .... **(1 mark)**

(iv) One termination step .... **(1 mark)**

(c) Several organic products are formed. Suggest how they may be separated.

.... **(1 mark)**

3   Methane and chlorine react together to produce chloroalkanes including CH_3Cl and CH_2Cl_2.

(a) Name these two products.

CH_3Cl .... CH_2Cl_2 ....   **(2 marks)**

(b) Write two equations to show the propagation steps in which CH_3Cl is converted into CH_2Cl_2.

....

.... **(2 marks)**

(c) Other chloroalkanes, including CH_3CH_2Cl, may also form.
Write an equation to represent a step in the mechanism in which this is the only product.

> You will need to show the reaction between two radicals.

.... **(1 mark)**

4   In chemistry, what is a radical?

.... **(1 mark)**

50

Alkenes and hydrogen halides

1　Ethene reacts with hydrogen bromide to form bromoethane:

$C_2H_4 + HBr \rightarrow CH_3CH_2Br$

What is the mechanism for this reaction?

☐　**A**　nucleophilic addition

☐　**B**　radical substitution

☐　**C**　electrophilic addition

☐　**D**　electrophilic substitution　　　　　　　　　　　　　　　　**(1 mark)**

2　Alkenes are hydrocarbons that contain a C=C bond. This comprises a π bond and a σ bond.

> Mention in your answer how s-orbitals overlap in σ bonds, how p-orbitals overlap in π bonds, and by how much in each case.

　　(a)　Suggest why the σ bond is stronger than the π bond.

　　..

　　.. **(2 marks)**

　　(b)　Draw a labelled diagram of ethene with its π and σ bond between two carbon atoms.

　　

　　　　　　　　　　　　　　　　　　　　　　　　　　　　　　　　(2 marks)

⟩Guided⟩　3　What is an electrophile?

　　An electrophile is a species that can .... **(1 mark)**

4　2-bromopropane, $CH_3CHBrCH_3$, is a product formed when propene reacts with hydrogen bromide.

　　(a)　Tick one box to indicate which carbocation leads to the formation of this product.　**(1 mark)**

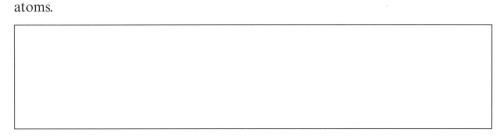

　　(b)　Complete the diagram below to show the mechanism for the reaction, showing the carbocation and product clearly.

> Use curly arrows to represent the movement of pairs of electrons.

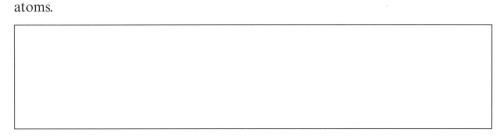

　　　　　　　　　　　　　　　　　　　　　　　　　　　　　　　　(4 marks)

More addition reactions of alkenes

1 (a) Tick three boxes to identify each species as a primary, secondary or tertiary carbocation.

(i) 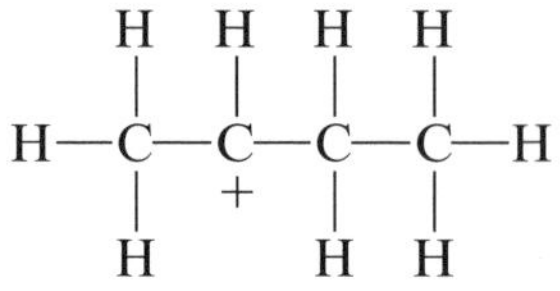   H H H H
      │ │ │ │
  H—C—C—C—C—H
      │ + │ │
      H   H H

(ii) 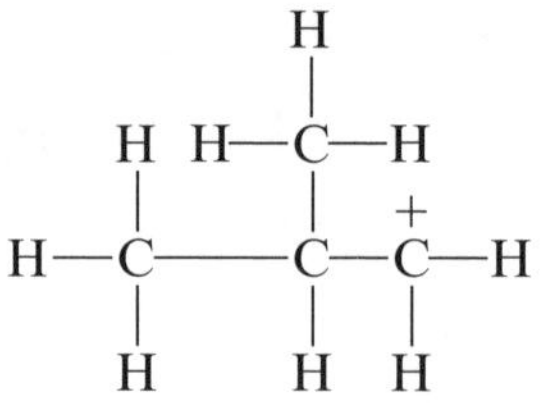   H
            │
     H H—C—H
     │   │ +
  H—C——————C—C—H
     │   │ │
     H   H H

(iii) 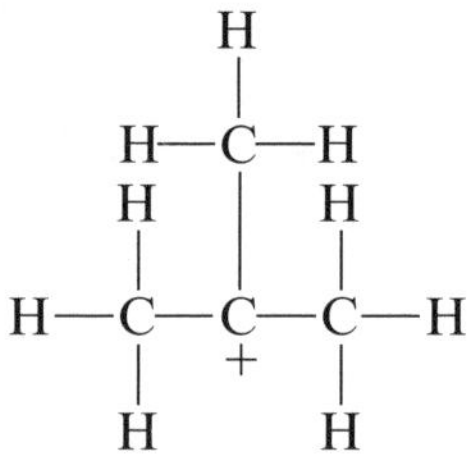   H
              │
          H—C—H
          H │ H
          │ │ │
     H—C—C—C—H
          │ + │
          H   H

Primary	☐	☐	☐
Secondary	☐	☐	☐
Tertiary	☐	☐	☐

(3 marks)

(b) Which species (i, ii or iii) is likely to be the most stable?

Species ....   **(1 mark)**

2 But-1-ene reacts with hydrogen in the presence of nickel. It also reacts with bromine.

(a) State the function of the nickel in the reaction with hydrogen.

.... **(1 mark)**

(b) Name the products formed in the reaction of but-1-ene with hydrogen, and with bromine.

Product with hydrogen ....

Product with bromine .... **(2 marks)**

(c) Explain how a dipole is induced in hydrogen and bromine molecules in these reactions.

....

.... **(2 marks)**

3 But-1-ene reacts with hydrogen bromide to form a minor product and a major product.

(a) Name the **minor** product formed in the reaction.

.... **(1 mark)**

(b) Complete the diagram below to show the mechanism for the reaction to form the **major** product, showing the carbocation and this product clearly.

> Use curly arrows to represent the movement of pairs of electrons.

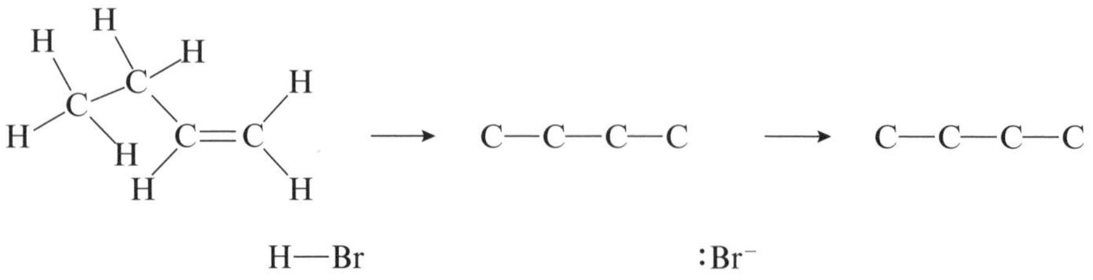

H   H
 \   │   H
  C—C   /
 / │ C═C
H   H   │ \
    H   H   H
         │
C—C—C—C → C—C—C—C

H—Br   :Br⁻

(5 marks)

Exam skills 5

1 1-chloropropane, $CH_3CH_2CH_2Cl$, can be made from propane by a free radical substitution reaction. It can also be made from propene by an electrophilic addition reaction.

(a) (i)   State the reagent and condition needed to make 1-chloropropane from propane.

....

.... **(2 marks)**

(ii) Write an equation for the initiation step.

.... **(1 mark)**

(iii) Write equations for the propagation steps.

....

.... **(2 marks)**

(iv) Explain, with the help of an equation, why hexane also forms.

....

.... **(2 marks)**

(v) Explain how 1-chloropropane may be separated from the reaction mixture.

....

.... **(2 marks)**

(b) (i)   Give the mechanism for the reaction between hydrogen chloride and propene to form the major product.

(3 marks)

(ii) State the type of mechanism involved.

.... **(1 mark)**

(iii) Explain why 1-chloropropane is the minor product in this reaction.

....

.... **(2 marks)**

Alkenes and alcohols

1 Ethanol can be made by reacting ethene with steam in the presence of phosphoric acid.

(a) Write an equation for this reaction.

.... **(1 mark)**

(b) Water is involved in this reaction. Name the type of reaction and the reaction mechanism.

Type of reaction ....

Reaction mechanism .... **(2 marks)**

(c) Ethene can be made by reacting ethanol at 180 °C in the presence of phosphoric acid. Name the type of reaction.

.... **(1 mark)**

(d) Why is phosphoric acid needed in these two reactions?

.... **(1 mark)**

2 Ethene can react with acidified potassium manganate(VII) solution, $KMnO_4$, to produce ethane-1,2-diol:

$$
\begin{array}{ccc}
 & H & H \\
 & | & | \\
H- & C- & C-H \\
 & | & | \\
 & OH & OH
\end{array}
$$

(a) This reaction could be classified as both:

☐ **A** radical substitution and oxidation

☐ **B** electrophilic addition and oxidation

☐ **C** radical substitution and reduction

☐ **D** nucleophilic addition and reduction. **(1 mark)**

(b) Propene can also react with acidified potassium manganate(VII) solution.

(i) Name the product formed in this reaction.

.... **(1 mark)**

(ii) Write an equation for the reaction.

....

....

> Show acidified potassium manganate(VII) solution as [O] in the equation.

(2 marks)

(iii) State the role of acidified potassium manganate(VII) solution in the reaction.

.... **(1 mark)**

(c) Name the acid used to acidify potassium manganate(VII) solution in these reactions.

> This acid cannot form chlorine when it is mixed with $KMnO_4$.

.... **(1 mark)**

Addition polymerisation

1 The diagram below shows a section of an addition polymer.

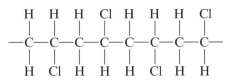

(a) In the boxes below, draw the displayed formulae for the polymer's repeat unit and monomer.

Both contain two carbon atoms.

repeat unit	monomer

(2 marks)

(b) State the systematic name for the monomer and for the polymer.

Monomer ....

Polymer .... **(2 marks)**

2 Propene forms poly(propene).
The structure of propene is shown on the right.

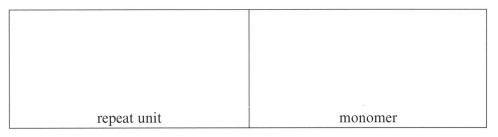

(a) In the boxes below, draw the displayed formulae for the repeat unit of poly(propene) and a short section of the polymer molecule.

repeat unit	section of polymer

(2 marks)

(b) The empirical formula of poly(propene) is the same as the empirical formula of propene.
Suggest why this is.

.... **(1 mark)**

(c) What is the atom economy for the production of poly(propene) from propene?
Give a reason for your answer.

.... **(1 mark)**

3 Describe what you expect to see if bromine water is added to ethene and to poly(ethene).
Explain your answer.

....

.... **(2 marks)**

Polymer waste

1   Waste polymers can be separated into different types for recycling, incineration or cracking.

   (a)   Explain what is meant by **incineration**.

   .... **(2 marks)**

   (b)   Incineration releases waste gases, including sulfur dioxide. This can be absorbed using calcium carbonate, $CaCO_3$. The reaction forms calcium sulfite, $CaSO_3$, and a gaseous product.

      (i)   Write an equation for the reaction.

      .... **(1 mark)**

      (ii)   State a problem caused by sulfur dioxide emissions.

      .... **(1 mark)**

2   Poly(lactic acid) is a biodegradable polymer made from lactic acid (2-hydroxypropanoic acid).

   (a)   Explain what is meant by the term *biodegradable*.

   ....

   .... **(2 marks)**

   (b)   Poly(lactic acid) can be converted back to its monomer using bacteria, enzymes or by chemical reactions at moderate temperatures. Addition polymers such as poly(propene) may be converted to a mixture similar to naphtha (an oil fraction) by cracking.
   Suggest two advantages of recycling poly(lactic acid) rather than poly(propene).

   ....

   .... **(2 marks)**

3   The table shows information about making a 'single use' poly(ethene) shopping bag, or a poly(ethene) 'bag for life'. Use the information to help you answer the questions.

Type of bag	Mass of raw material /g	Energy used to make it /kJ
'Single use' bag	8	20
'Bag for life'	32	160

   (a)   Suggest an advantage and a disadvantage of 'single use' bags.

   ....

   .... **(2 marks)**

   (b)   To what extent is the 'bag for life' a more sustainable use of materials and energy?

   Compare the amounts of raw materials and energy to make each bag, and write a conclusion.

   ....

   ....

   .... **(3 marks)**

Alcohols from halogenoalkanes

1   Bromoethane reacts with aqueous potassium hydroxide solution under reflux to form ethanol:

$$CH_3CH_2Br + OH^- \rightarrow CH_3CH_2OH + Br^-$$

What is the mechanism for this reaction?

☐ **A** nucleophilic addition

☐ **B** radical substitution

☐ **C** electrophilic addition

☐ **D** nucleophilic substitution   **(1 mark)**

2   Name the following compounds.

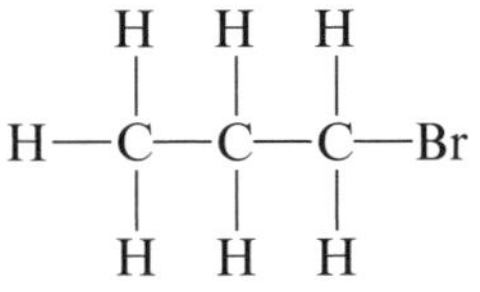

(a) 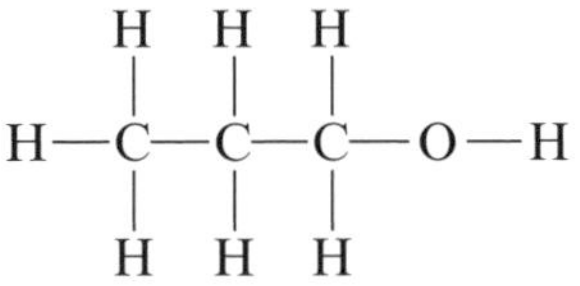 ....   (b) ....

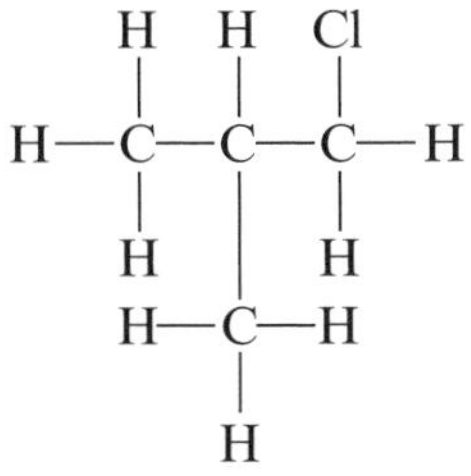

(c) 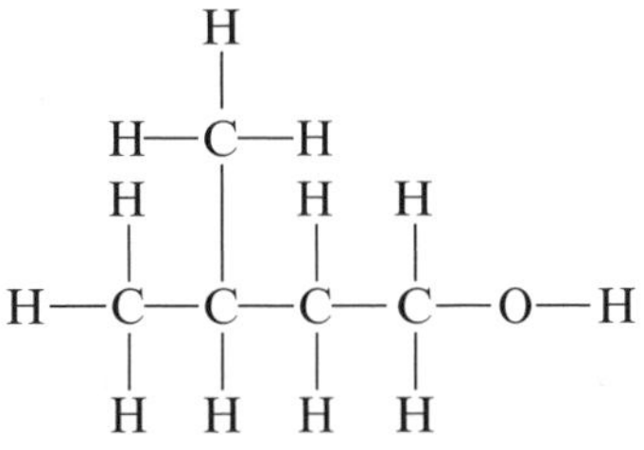 ....   (d) ....   **(4 marks)**

⟩**Guided**⟩   3   What is a **nucleophile**?

A nucleophile is a species that can ....   **(1 mark)**

4   State why reflux is used in the preparation of organic compounds, and describe how it works.

....

....

....   **(3 marks)**

5   1-bromobutane reacts with potassium hydroxide solution under reflux to produce butan-1-ol.

(a)   Complete the diagram below to show the mechanism for the reaction, showing the reactants and products clearly.

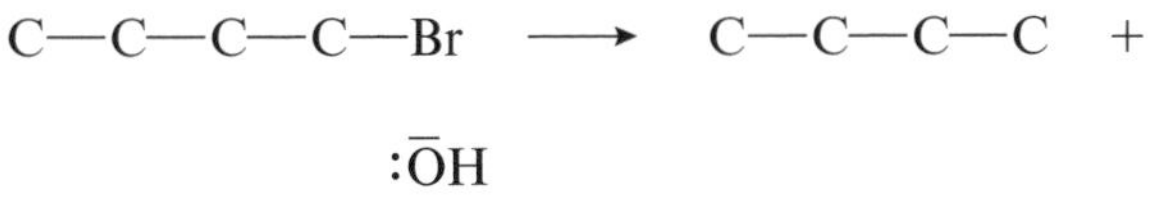

C—C—C—C—Br   ⟶   C—C—C—C   +

:ŌH

Use curly arrows to represent the movement of pairs of electrons.   **(5 marks)**

(b)   State the role of the K^+ ion, and the role of OH^- ion, in this reaction.

K^+ ion ....

OH^- ion ....   **(2 marks)**

Reactivity of halogenoalkanes

1 (a) Tick three boxes to identify each compound as a **primary**, **secondary** or **tertiary** chloroalkane.

(i)

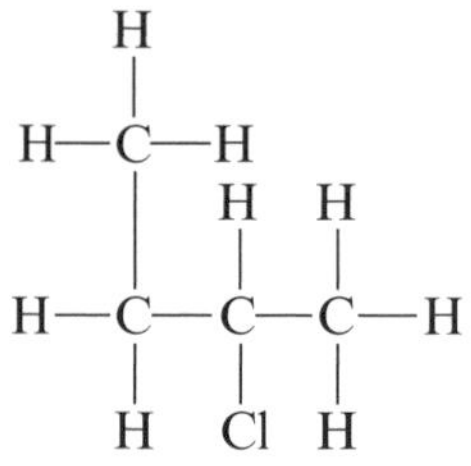

```
        H
        |
    H—C—H
        H   H
        |   |
H—C—C—C—H
    |   |   |
    H   Cl  H
```

(ii)

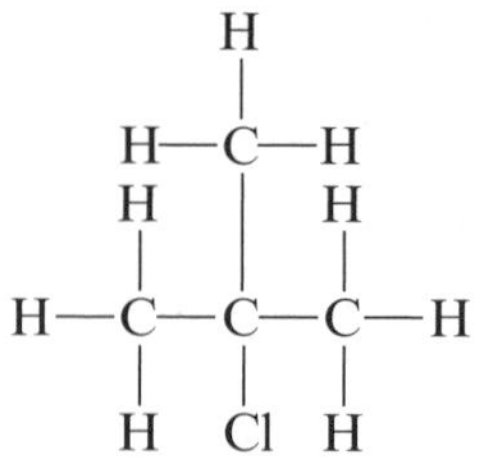

```
        H
        |
    H—C—H
    H   |   H
    |   |   |
H—C—C—C—H
    |   |   |
    H   Cl  H
```

(iii)

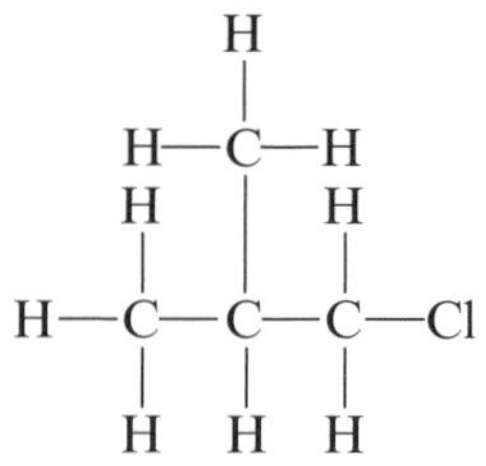

```
        H
        |
    H—C—H
    H   |   H
    |   |   |
H—C—C—C—Cl
    |   |   |
    H   H   H
```

Primary   ☐   ☐   ☐
Secondary   ☐   ☐   ☐
Tertiary   ☐   ☐   ☐

(3 marks)

(b) Which compound (i, ii or iii) is likely to be the most reactive?

Compound .... **(1 mark)**

> Guided >

2 What is a meant by the term **hydrolysis**?

A reaction in which .... **(1 mark)**

3 1-bromopropane reacts with water to produce an alcohol and an inorganic product.

(a) Write an equation to represent this reaction, and name the organic product.

Equation: ....

Name of organic product: .... **(2 marks)**

(b) State the type of reaction mechanism involved.

.... **(1 mark)**

4 Silver nitrate solution can be used to identify halide ions in solution.

(a) State the colours of the precipitates formed in the reaction between Ag^+ ions and:

Cl^- ions .... Br^- ions .... I^- ions .... **(1 mark)**

(b) These reactions can be used to investigate the rate of hydrolysis of halogenoalkanes. Explain why ethanol is added to the reaction mixture.

.... **(1 mark)**

5 Describe, and explain in terms of bond enthalpy, the differences in reactivity of chloroethane, bromoethane and iodoethane.

....

....

....

.... **(4 marks)**

More halogenoalkane reactions

1 Which of the following is **not** a correct name for the compound on the right?

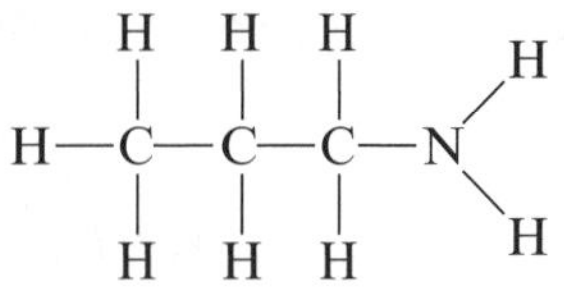

H—C—C—C—N (with H atoms: H H H on top, H H H on bottom, and two H on the N)

☐ **A** propan-1-amine

☐ **B** propylamine

☐ **C** 1-aminopropane

☐ **D** prop-1-amino

(1 mark)

2 1-bromopropane reacts with excess ammonia to form the compound shown in question 1.

> Complete the central organic structure, and use curly arrows to represent the movement of pairs of electrons.

(a) Complete the diagram below to show the mechanism for the reaction.

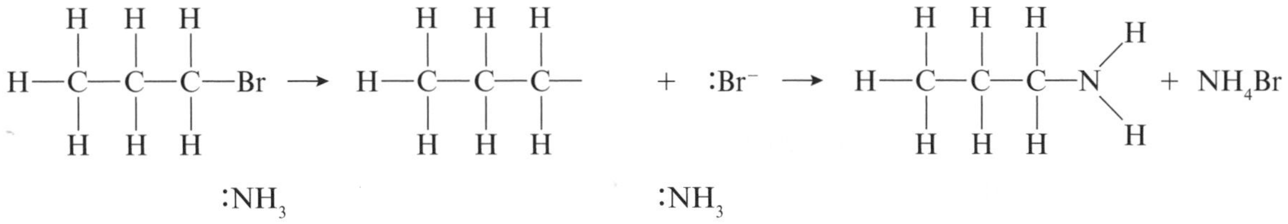

H—C—C—C—Br → H—C—C—C—   + :Br⁻ → H—C—C—C—N + NH_4Br

:NH₃   :NH₃

(5 marks)

(b) State the roles of the first and second ammonia molecules in this reaction mechanism.

In the first step, NH_3 acts as ....

In the second step, NH_3 acts as ....

(2 marks)

3 1-bromopropane reacts, under reflux, with potassium cyanide dissolved in ethanol.

(a) Write an equation to represent this reaction, and name the organic product.

Equation: ....

Name of organic product: ....

(2 marks)

(b) Identify the nucleophile in this reaction.

....

(1 mark)

4 2-bromopropane reacts, under reflux, with potassium hydroxide dissolved in ethanol.

(a) Write an equation to represent this reaction, and name the organic product.

Equation: ....

Name of organic product: ....

(2 marks)

(b) State the type of reaction this is, and the role of the hydroxide ion.

Type of reaction: ....

Role of OH⁻ ion: ....

(2 marks)

Oxidation of alcohols

1 Tick three boxes to identify each compound as a primary, secondary or tertiary alcohol.

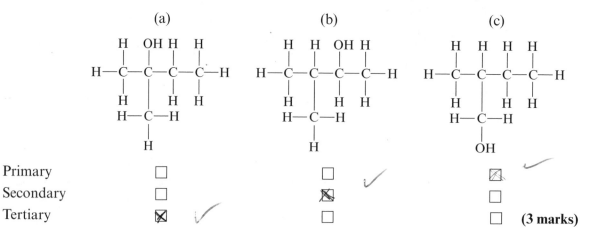

	(a)	(b)	(c)
Primary	☐	☐ ✓	☒ ✓
Secondary	☐	☒	☐
Tertiary	☒ ✓	☐	☐ **(3 marks)**

2 Ethanol is mixed with petrol for use as a fuel for cars.

(a) Write an equation to represent the complete combustion of ethanol.

$C_2H_5OH + 3O_2 \rightarrow 2CO_2 + 3H_2O$ ✓ **(1 mark)**

(b) Write an equation to represent the incomplete combustion of ethanol to form water, and equal amounts of carbon monoxide and a solid product.

$C_2H_5OH + \frac{3}{2}O_2 \rightarrow CO + 3H_2O + C$ **(1 mark)**

3 Ethanol is oxidised when it reacts with potassium dichromate(VI), acidified with sulfuric acid.

> Represent the oxidising agent as [O], rather than writing its formula.

(a) Write an equation to represent the reaction under distillation, and name the organic product.

Equation: $C_2H_5OH + [O] \rightarrow C_2H_4O + H_2O$ ✓

Name of organic product: Ethanal **(2 marks)**

(b) Write an equation to represent the reaction under reflux, and name the organic product.

Equation: $C_2H_5OH + 2[O] \rightarrow CH_3COOH + H_2O$

Name of organic product: Ethanoic acid **(2 marks)**

4 Propan-1-ol and propan-2-ol, are oxidised by acidified potassium dichromate(VI) under distillation.

(a) Name the organic product of each reaction.

From propan-1-ol: propanal ✓

From propan-2-ol: propanone ✓ **(2 marks)**

(b) Describe a simple laboratory test to distinguish between these two products.

Reaction with tollen's reagent ; ketone won't react, aldehyde will form silver mirror ✓ **(2 marks)**

Halogenoalkanes from alcohols

1   Chloroalkanes can be prepared from alcohols.

(a)   Write an equation to represent the reaction between propan-1-ol and phosphorus(V) chloride.

> Two inorganic products are also made in the reaction.

$C_3H_5OH + PCl_5 \rightarrow$ .... **(2 marks)**

(b)   Suggest a method to separate the organic product from the reaction mixture.

.... **(1 mark)**

(c)   Describe how you could determine the purity of this separated product.

....

.... **(2 marks)**

2   Bromoalkanes can be prepared from alcohols.

(a)   State the two inorganic reagents needed to produce bromoalkanes from alcohols.

.... **(1 mark)**

(b)   The mixture of reagents in part (a) produces hydrogen bromide.
Write an equation to represent the reaction between this and butan-1-ol.

.... **(1 mark)**

3   Iodoalkanes can be prepared from alcohols.

(a)   State the two inorganic reagents needed to produce iodoalkanes from alcohols.

.... **(1 mark)**

(b)   The mixture of reagents in part (a) produces phosphorus(III) iodide.
Write an equation to represent the reaction between this and propan-1-ol.

.... **(2 marks)**

4   Concentrated hydrochloric acid reacts with 2-methylpropan-2-ol to form
2-chloro-2-methylpropane. The reaction can be carried out in a separating funnel.
The upper layer contains the organic product, so the lower layer is run off and discarded.

(a)   Suggest why two layers form in the separating funnel.

.... **(1 mark)**

(b)   The separated upper layer can be dried to remove water that may be present.
Name a suitable anhydrous salt that could be used as the drying agent.

.... **(1 mark)**

Exam skills 6

1 (a) Propanoic acid can be made from propan-1-ol by reaction with excess potassium dichromate(VI) solution, acidified with dilute sulfuric acid.

   (i) Write an equation for the reaction.

   .... **(1 mark)**

   (ii) Draw a diagram to show the laboratory apparatus needed to make propanoic acid from propan-1-ol.

   (3 marks)

   (iii) Describe the colour change you would expect to see in the reaction.

   .... **(1 mark)**

   (iv) When the final reaction mixture is distilled, an aqueous solution of propanoic acid is produced. State why the excess reactants are left behind.

   .... **(1 mark)**

   (v) Suggest how you could produce pure propanoic acid from the solution of propanoic acid.

   .... **(1 mark)**

   (b) 1-iodopropane can be made by reacting propan-1-ol with phosphorus(III) iodide, PI_3.

   (i) Write an equation to show the formation of phosphorus(III) iodide from moist red phosphorus and iodine.

   .... **(1 mark)**

   (ii) Write an equation to show the reaction between propan-1-ol and phosphorus(III) iodide.

   .... **(1 mark)**

   (iii) 1-iodopropane reacts with hot, aqueous silver nitrate solution. Describe what you would see when this reaction happens.

   .... **(1 mark)**

Structures from mass spectra

1   The diagram shows the mass spectrum for a compound of carbon, hydrogen and oxygen.

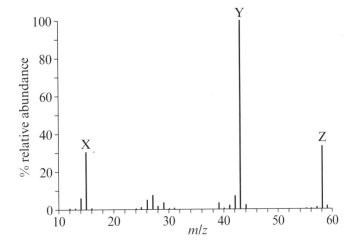

(a)   State the m/z ratios of the peaks labelled X, Y and Z.

Peak X: ....

Peak Y: ....

Peak Z: ....   **(1 mark)**

(b)   Identify the molecular ion peak and so determine the compound's relative molecular mass.

.... **(1 mark)**

(c)   Suggest the identity of the species responsible for the peak labelled X.

....   **(2 marks)**

(d)   The species responsible for the peak labelled Y contains one oxygen atom.

(i)   Suggest the identity of this species.

....   **(2 marks)**

(ii)   State the structural formula for the compound, and its name.

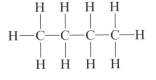

Look at your answers to parts (c) and (d)(i).

.... **(2 marks)**

2   Butane and methylpropane, shown, are chain isomers of C_4H_{10}. At which of the following m/z ratios would you expect there be a major peak in the mass spectrum of butane, but not in the mass spectrum of methylpropane?

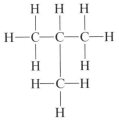

☐ **A**   15

☐ **B**   29

☐ **C**   43

☐ **D**   58   **(1 mark)**

Infrared spectroscopy

1 What effect does infrared radiation have on the N−H bonds in amines such as CH_3NH_2?

☐ **A** It has no effect because N−H bonds are not polar.

☐ **B** Heterolytic fission occurs, producing ions.

☐ **C** Homolytic fission occurs, producing radicals.

☐ **D** The bonds vibrate more vigorously.

(1 mark)

Bond	C−H	C=C	C=O	O−H (alcohols)	O−H (carboxylic acids)
Wavenumber range /cm^{-1}	3095−2700	1669−1645	1740−1700	3750−3200	3300−2500

2 The diagram is an infrared spectrum of an organic compound.

> Use these wavenumbers to help you to answer the questions below.

(a) The compound is most likely to be:

☐ **A** an alcohol

☐ **B** a carboxylic acid

☐ **C** an aldehyde

☐ **D** an alkene.

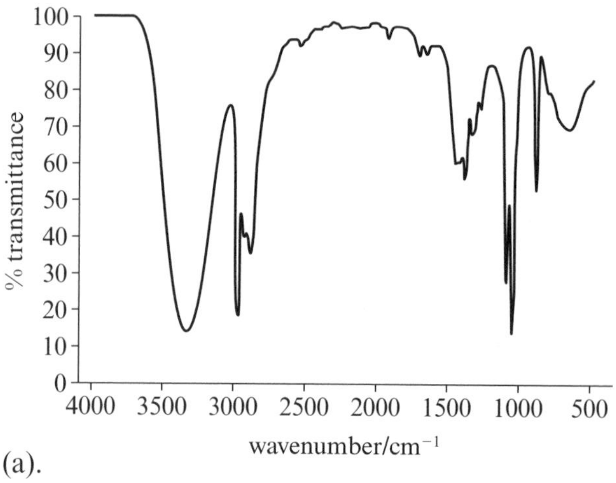

(1 mark)

(b) Explain your answer to part (a).

....

....

.... **(3 marks)**

3 Ethene reacts with steam to form a single product.

(a) Describe a peak seen in the infrared spectrum of ethene, but not in the product.

.... **(1 mark)**

(b) Describe a peak seen in the infrared spectrum of the product, but not in ethene.

.... **(1 mark)**

4 Propan-2-ol is oxidised by acidified potassium dichromate(VI).

(a) Name the organic product of this reaction.

.... **(1 mark)**

(b) Describe a peak seen in the infrared spectrum of the product, but not in propan-2-ol.

.... **(1 mark)**

Enthalpy changes

1. In experiments to measure the enthalpy change of reactions involving gases, which of the following conditions must always be kept constant?

   ☒ **A** temperature and pressure

   ☐ **B** temperature   *c*

   ☐ **C** pressure

   ☐ **D** volume   **(1 mark)**

2. Standard enthalpy changes are measured under standard conditions.

   (a) State the standard conditions of temperature and of pressure.

   Temperature: .... 298 k ✓ ....

   Pressure: .... 100 kPa. ✓ .... **(2 marks)**

   > **Guided**

   (b) Define the term **standard enthalpy change of formation, ΔH°**.

   This is the enthalpy change when .... I mol of a product

   ...... is formed from elements of the

   ...... product in their standard state. ...... **(3 marks)**

   (c) Write an equation to represent the standard enthalpy change of formation of sodium bromate(V), $NaBrO_3(s)$. Include state symbols in your answer.

   .... $2Na + Br_2 + 3O_2 \longrightarrow 2NaBrO_3$ .... **(2 marks)**

3. Complete the table to show the temperature changes, and signs of the enthalpy changes, for exothermic and endothermic reactions in solution.

Type of reaction	Temperature change	Sign of ΔH (+ or −)
Exothermic		—
Endothermic		+

   (2 marks)

4. Baking powder contains sodium hydrogencarbonate. This decomposes in the heat of an oven:

   $$2NaHCO_3(s) \rightarrow Na_2CO_3(s) + H_2O(l) + CO_2(g) \qquad \Delta H = +91.6\,kJ\,mol^{-1}$$

   In the space below, draw a labelled enthalpy level diagram for this reaction.

   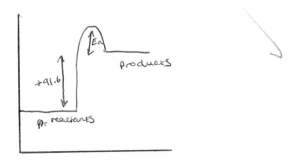

   (3 marks)

Measuring enthalpy changes

1 The enthalpy change of combustion of an alcohol can be determined by **calorimetry**.
A known volume of water was heated using a spirit burner of alcohol, and the increase in
temperature measured. The mass of the spirit burner was measured before and after heating.

When the results were used to calculate the enthalpy change, it was less than the accepted value.

Which of these factors is most likely to cause this difference?

☐ **A** Using a ±0.1 g balance rather than a ±0.01 g balance.

☐ **B** Using a measuring cylinder instead of a pipette for the water.

☒ **C** Evaporation of alcohol during the experiment.

☐ **D** Heat losses from the water and its container.

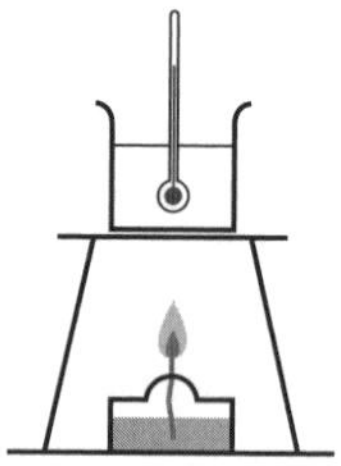

(1 mark)

2 The following results were obtained in an experiment, like the one in question 1, to measure the
enthalpy change of combustion of methanol. In the experiment, 150.0 g of water was used.

Initial temperature of water /°C	Final temperature of water /°C	Initial mass of spirit burner /g	Final mass of spirit burner /g
21.8	40.3	123.64	122.78

(a) Calculate the heat energy produced by the combustion of the methanol.

> heat energy (J) = mass of water (g) × 4.18 × temperature change (K)

$$40.3$$
$$21.8$$
$$18.5$$

150 × 4.18 × 18.5 = 11 599.5 ✓ ✓

(2 marks)

(b) Calculate the amount in moles of methanol, CH_3OH, burned. Give your answer
to three significant figures.

12 + 4 + 16 = 32.

0.86 g

$$\frac{0.86}{32} = 0.026875$$

(3 marks)

(c) Use your answers to parts (a) and (b) to calculate $\Delta_c H$ for methanol in $kJ\,mol^{-1}$.
Give your answer the appropriate sign and to three significant figures.

$$\frac{11.599\ kJ}{0.026875} = 431.590697\ 7$$

(3 marks)

(d) State the assumption made in this experiment and the calculations in part (c).

The temp and pressure were constant

(1 mark)

3 Define the term **standard enthalpy change of neutralisation**, $\Delta_{neut}H^\ominus$.

(3 marks)

Enthalpy cycles

⟩**Guided**⟩ **1** State Hess's Law.

The enthalpy change for a reaction is .... *the* .... *change* .... *in* ....

.... *energy* .... *between* .... *the* .... *reactants* .... *and* .... **(2 marks)**
products.

2 The enthalpy change of formation for ethene cannot be measured directly, but this enthalpy cycle can be used to determine it indirectly.

$$3O_2(g) + 2C(S) + 2H_2(g) \xrightarrow{\Delta_f H} CH_2{=}CH_2(g) + 3O_2(g)$$
$$\Delta_A H \searrow \qquad \swarrow \Delta_B H$$
$$2CO_2(g) + 2H_2O(l)$$

The enthalpy change of formation, $\Delta_f H$, is equal to:

☐ **A** $\Delta H_A + \Delta H_B$

☒ **B** $\Delta H_A - \Delta H_B$

☐ **C** $-\Delta H_A + \Delta H_B$

☐ **D** $-\Delta H_A - \Delta H_B$ **(1 mark)**

3 The enthalpy change of formation of methanol, $CH_3OH(l)$, is difficult to measure directly.

(a) Write the chemical equation that represents $\Delta_f H$ of methanol. Include state symbols.

.... $C + 2H_2 + \frac{1}{2}O_2 \longrightarrow CH_3OH$ .... **(1 mark)**

(b) These equations can be combined to form an enthalpy cycle for $\Delta_f H[CH_3OH(l)]$:

- $C(s) + 2H_2(g) + 2O_2(g) \rightarrow CO_2(g) + 2H_2O(l)$
- $CH_3OH(l) + 1\frac{1}{2}O_2(g) \rightarrow CO_2(g) + 2H_2O(l)$
- $C(s) + 2H_2(g) + 2O_2(g) \rightarrow CH_3OH(l) + 1\frac{1}{2}O_2(g)$

In the space below, use these equations to draw a suitable enthalpy cycle.

$$C + 2H_2 + \frac{1}{2}O_2 \xrightarrow{\Delta_f H} CH_3OH.$$
$$\searrow \qquad \nwarrow$$
$$CO_2 + H_2O.$$

(1 mark)

4 Ammonia reacts with hydrogen chloride to form ammonium chloride:

$$NH_3(g) + HCl(g) \rightarrow NH_4Cl(s)$$

In the space below, draw an enthalpy cycle that could be used to determine the enthalpy change of this reaction, $\Delta_r H$, using enthalpy change of formation data. **(1 mark)**

$$NH_3 + HCl \longrightarrow NH_4Cl$$

$$N + H_2 + Cl_2$$

Using enthalpy cycles

1 In general, the standard enthalpy change for a reaction, $\Delta_r H^\circ$, refers to: reactants → products.

$\Delta_r H^\circ$ may be calculated using standard enthalpy changes of formation, $\Delta_f H^\circ$, for the reactants and products, or using standard enthalpy changes of combustion, $\Delta_c H^\circ$ for the reactants and products.

Explain, with the help of an equation, how you can calculate $\Delta_r H^\circ$ using $\Delta_f H^\circ$ data or $\Delta_c H^\circ$ data.

Using $\Delta_f H^\circ$ data: .... **(1 mark)**

Using ΔH° data: .... **(1 mark)**

2 Explain why the value for $\Delta H_f^\circ[CO_2(g)]$ is identical to the value for $\Delta H_c^\circ[C(s)]$.

> Think about the equation that represents the chemical reaction for each change.

.... **(1 mark)**

3 The data in the table can be used in the calculation of the standard enthalpy change of combustion of methane:
$CH_4(g) + 2O_2(g) \rightarrow CO_2(g) + 2H_2O(l)$

Substance	$CH_4(g)$	$CO_2(g)$	$H_2O(l)$
$\Delta_f H^\circ$/kJ mol^{-1}	−75	−394	−286

(a) Explain why a value for $\Delta_f H^\circ[O_2(g)]$ is not given in the table.

.... The O_2 itself   doesn't   combust .... **(1 mark)**

(b) Calculate the standard enthalpy change of combustion of methane, $\Delta_c H^\circ[CH_4(g)]$.

....

....

.... **(3 marks)**

4 The data in the table can be used in the calculation of the standard enthalpy change of formation of ethanol:

Substance	C(s)	$H_2(g)$	$CH_3CH_2OH(l)$
$\Delta_c H^\circ$/kJ mol^{-1}	−394	−286	−1367

$2C(s) + 3H_2(g) + \frac{1}{2}O_2(g) \rightarrow CH_3CH_2OH(l)$

Calculate the standard enthalpy change of formation of ethanol, $\Delta_f H^\circ[CH_3CH_2OH(l)]$.

....

....

.... **(3 marks)**

5 $\Delta_f H^\circ[C_2H_4(g)] = +52\,kJ\,mol^{-1}$ and $\Delta_f H^\circ[C_2H_6(g)] = -85\,kJ\,mol^{-1}$
What is the standard enthalpy change of reaction, $\Delta_r H^\circ$, for the reaction:
$C_2H_4(g) + H_2(g) \rightarrow C_2H_6(g)$?

☐ **A** $-33\,kJ\,mol^{-1}$

☐ **B** $+33\,kJ\,mol^{-1}$

☐ **C** $-137\,kJ\,mol^{-1}$

☐ **D** $+137\,kJ\,mol^{-1}$ **(1 mark)**

Mean bond enthalpy calculations

Guided

1 Define the term **bond enthalpy**.

It is the enthalpy change when .... I mol of bonds are ....

.... broken . ... **(2 marks)**

2 Hydrogen reacts with oxygen to form water:

$$[H-H] + \tfrac{1}{2}[O=O] \rightarrow [H-O-H]$$

Bond	H–H	O=O	O–H
Bond enthalpy /kJ mol^{-1}	436	498	464

(a) Use the data in the table to calculate the enthalpy change for this reaction.

....

....

.... **(3 marks)**

(b) The data book value for the standard enthalpy change of combustion, $\Delta_c H^\ominus$, for hydrogen is $-286\,\text{kJ mol}^{-1}$. Explain why the value calculated in part (a) is less exothermic than this value.

> The values quoted in the table are not mean bond enthalpies.

.... **(1 mark)**

3 Which of the following equations represents the reaction for which the enthalpy change, ΔH, is the mean bond enthalpy of the C−Cl bond?

☐ **A** $\tfrac{1}{4}CCl_4(g) \rightarrow \tfrac{1}{4}C(g) + Cl(g)$

☐ **B** $\tfrac{1}{4}C(g) + Cl(g) \rightarrow \tfrac{1}{4}CCl_4(g)$

☐ **C** $CCl_4(g) \rightarrow C(g) + 4Cl(g)$

☐ **D** $C(g) + 4Cl(g) \rightarrow CCl_4(g)$ **(1 mark)**

4 Ethene reacts with hydrogen chloride to form chloroethane:

$$CH_2=CH_2(g) + HCl(g) \rightarrow CH_3CH_2Cl(g)$$

The standard enthalpy change for this reaction, $\Delta_f H^\ominus$, is $-97\,\text{kJ mol}^{-1}$.

(a) Use this value, and the data in the table, to calculate the bond enthalpy for the H−Cl bond.

Bond	C–C	C=C	C–H	C–Cl
Mean bond enthalpy /kJ mol^{-1}	347	612	413	346

....

....

....

.... **(3 marks)**

(b) A data book gives a value of $432\,\text{kJ mol}^{-1}$ for the H−Cl bond. Explain the difference between this value and the one calculated in part (a).

....

.... **(1 mark)**

Changing reaction rate

1 Copper(II) oxide reacts with dilute sulfuric acid to form copper(II) sulfate and water:

$$CuO(s) + H_2SO_4(aq) \rightarrow CuSO_4(aq) + H_2O(l)$$

Which of these factors will **not** affect the rate of this reaction?

☐ **A** concentration

☐ **B** temperature

☐ **C** surface area

☐ **D** pressure **(1 mark)**

2 State the meaning of the term **rate of reaction**.

....

.... **(2 marks)**

3 Factors that affect the rate of chemical reactions can be understood using collision theory and considerations of the energy involved.

(a) What is meant by the term **activation energy**?

.... **(1 mark)**

(b) State one factor, other than energy, that may determine whether a collision will be successful.

.... **(1 mark)**

4 Calcium carbonate reacts with dilute hydrochloric acid:

$$CaCO_3(s) + 2HCl(aq) \rightarrow CaCl_2(aq) + H_2O(l) + CO_2(g)$$

(a) What is the effect on the reaction rate of using powdered calcium carbonate instead of lumps?

.... **(1 mark)**

(b) Explain, in terms of collisions, your answer to part (a).

....

.... **(2 marks)**

5 Sodium thiosulfate reacts with dilute hydrochloric acid:

$$Na_2S_2O_3(aq) + 2HCl(aq) \rightarrow 2NaCl(aq) + H_2O(l) + SO_2(g) + S(s)$$

(a) What is the effect on the reaction rate of increasing the concentration of hydrochloric acid?

.... **(1 mark)**

(b) Explain, in terms of collisions, your answer to part (a).

....

.... **(2 marks)**

Maxwell–Boltzmann model

1 The diagram shows the Maxwell–Boltzmann distribution of molecular energies in a sample of gas at a temperature, T_1. The energy marked E_a is the activation energy for a reaction involving the gas in the sample.

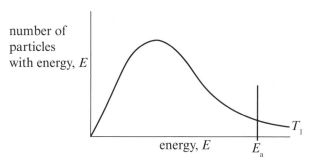

(a) Energy, E, is plotted on the horizontal axis. This energy is mainly:

    ☐ **A** vibration

    ☐ **B** rotation

    ☐ **C** kinetic

    ☐ **D** activation   **(1 mark)**

(b) Mark a cross, ✕, on one axis of the graph to show the most probable energy.   **(1 mark)**

> Draw a horizontal or vertical line from the curve to one of the axes, and mark ✕ where it crosses that axis.

(c) What does the area under the curve represent?

.... **(1 mark)**

(d) Shade a part of the diagram to represent the molecules with sufficient energy to react.   **(1 mark)**

(e) Draw a distribution, on the diagram above, to represent the same sample of gas but at a higher temperature. Mark your line, T_2.   **(2 marks)**

(f) What happens to the most probable energy, and the number of molecules with this energy, as the temperature of the gas increases?

....

.... **(2 marks)**

(g) Explain why an increase in temperature leads to an increase the rate of reaction. Refer to the Maxwell–Boltzmann distribution in your answer.

....

....

.... **(2 marks)**

Catalysts

1 The diagram shows the Maxwell–Boltzmann distribution of molecular energies for an uncatalysed reaction.

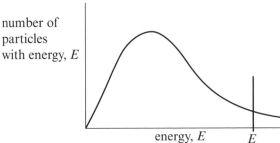

Which of the following would move the activation energy, E_a, to the left?

☐ **A** adding a catalyst

☐ **B** increasing the temperature

☐ **C** reducing the reactant concentration

☐ **D** removing the product as it forms **(1 mark)**

Guided **2** Explain, in general terms, how a catalyst works.

A catalyst provides ....

....

> Details about homogeneous and heterogeneous catalysis are not required here.

(2 marks)

3 The reaction between hydrochloric acid and sodium hydroxide solution is exothermic:

$$HCl(aq) + NaOH(aq) \rightarrow NaCl(aq) + H_2O(l)$$

Complete the reaction profile diagram to represent the enthalpy changes in this reaction. On your diagram, clearly indicate the activation energy, E_a, and the enthalpy change, ΔH. **(4 marks)**

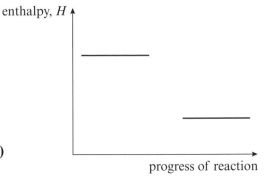

4 The decomposition of ammonia to form nitrogen and hydrogen is endothermic:

$$2NH_3(g) \rightarrow N_2(g) + 3H_2(g)$$

If a suitable cheap catalyst can be developed, ammonia could be used as a source of hydrogen for hydrogen–oxygen fuel cells in cars and other vehicles.

(a) On the diagram, draw the reaction profile for an uncatalysed endothermic reaction. Label this curve, **A**. **(1 mark)**

(b) Add to your diagram the reaction profile for the same reaction as in part (a), but in the presence of a catalyst. Label this curve, **B**. **(1 mark)**

(c) On your diagram, clearly indicate the activation energy, E_a, of the catalysed reaction. **(1 mark)**

Dynamic equilibrium 1

1 State the meaning of **dynamic** in the term **dynamic equilibrium**.

.... **(1 mark)**

2 Which of the following statements about the concentrations of reactants and products at equilibrium is always true?

   ☐ **A** The concentrations of reactants and products are the same.

   ☐ **B** The concentration of reactants is greater than the concentration of products.

   ☐ **C** The concentration of products is greater than the concentration of reactants.

   ☐ **D** The concentrations of reactants and products are constant.   **(1 mark)**

3 State a feature of the rates of the forward reaction and backward reaction at equilibrium.

.... **(1 mark)**

4 Ammonia is manufactured from nitrogen and hydrogen:

$$N_2(g) + 3H_2(g) \rightleftharpoons 2NH_3(g)$$

The graph shows how the equilibrium yield of ammonia varies with pressure at constant temperature.

(a) Describe the effect of pressure on the equilibrium yield of ammonia.

.... **(1 mark)**

(b) Explain the effect of reducing the pressure on the equilibrium yield of ammonia.

....

.... **(2 marks)**

(c) In the Haber process, the ammonia produced is cooled to liquefy it.
Predict and explain the effect on the position of equilibrium of removing ammonia from the equilibrium mixture.

....

.... **(2 marks)**

5 Ethanoic acid partially dissociates in aqueous solution:

$$CH_3COOH(aq) \rightleftharpoons CH_3COO^-(aq) + H^+(aq)$$

Predict and explain the effect on the position of equilibrium of dissolving some sodium ethanoate (a source of CH_3COO^- ions) in dilute ethanoic acid.

....

.... **(2 marks)**

Dynamic equilibrium 2

1 What effect does adding a catalyst have on the equilibrium position of a reversible reaction?

☐ **A** The forward reaction rate increases, so the position of equilibrium moves to the right.

☐ **B** The backward reaction rate decreases, so the position of equilibrium moves to the right.

☐ **C** The backward reaction rate increases, so the position of equilibrium moves to the left.

☐ **D** The rates of the forward and backward reaction increase by the same ratio, so the position of equilibrium stays the same. **(1 mark)**

2 The decomposition of hydrogen iodide is a reversible reaction:

$$2HI(g) \rightleftharpoons H_2(g) + I_2(g) \qquad \Delta_r H^\ominus = +9.4 \, kJ \, mol^{-1} \text{ (for the forward reaction)}$$

(a) State the standard enthalpy change of reaction for the backward reaction.

.... **(1 mark)**

(b) Predict and explain the effect on the position of equilibrium of increasing the temperature.

....

.... **(2 marks)**

3 (a) Write an expression for the equilibrium constant, K_c, for this general reversible reaction:

$$aA(aq) + bB(aq) \rightleftharpoons cC(aq) + dD(aq)$$

.... **(1 mark)**

(b) Write an expression for K_c for this reaction:

$$2SO_2(g) + O_2(g) \rightleftharpoons 2SO_3(g)$$

.... **(1 mark)**

⟩**Guided**⟩ **4** Explain the difference between a homogeneous system and a heterogeneous system.

In a homogeneous system the components are ....

.... **(1 mark)**

5 What can be said about the concentration of a solid or a liquid at a fixed temperature?

.... **(1 mark)**

6 Write an expression for the equilibrium constant, K_c, for these reversible reactions:

(a) $SO_2(g) + \frac{1}{2}O_2(g) \rightleftharpoons SO_3(g)$ $K_c = $ .... **(1 mark)**

(b) $2Ag^+(aq) + Cu(s) \rightleftharpoons 2Ag(s) + Cu^{2+}(aq)$ $K_c = $ .... **(1 mark)**

(c) $NH_4Cl(s) \rightleftharpoons NH_3(g) + HCl(g)$ $K_c = $ .... **(1 mark)**

(d) $Cl_2(aq) + H_2O(l) \rightleftharpoons HCl(aq) + HClO(aq)$ $K_c = $ .... **(1 mark)**

Industrial processes

1 Nitrogen and hydrogen react to form ammonia:

$$N_2(g) + 3H_2(g) \rightleftharpoons 2NH_3(g) \qquad \Delta H = -92 \, kJ \, mol^{-1}$$

What effect would increasing the temperature have on this system?

☐ **A** Yield of ammonia reduced but rate increased.

☐ **B** Yield of ammonia reduced and rate decreased.

☐ **C** Yield of ammonia increased but rate decreased.

☐ **D** Yield of ammonia increased and rate increased.   **(1 mark)**

2 In some processes, increasing the pressure increases the equilibrium yield.
  Give two reasons why, in practice, very high pressures may not be used in such processes.

....

.... **(2 marks)**

3 Ethene reacts with steam to produce ethanol:

$$C_2H_4(g) + H_2O(g) \rightleftharpoons CH_3CH_2OH(g) \qquad \Delta H = -49 \, kJ \, mol^{-1}$$

> Look back at page 54 for information about alcohols from alkenes.

(a) Identify a suitable catalyst for this reaction.

.... **(1 mark)**

(b) Predict the effect of increasing the temperature on the equilibrium yield of ethanol.

.... **(1 mark)**

(c) Considering your answer to part (b), suggest why a temperature of about 300 °C is used.

....

.... **(2 marks)**

(d) Give two ways, without altering the temperature, to increase the equilibrium yield of ethanol.

....

.... **(2 marks)**

4 One step in the manufacture of sulfuric acid involves this reaction:

$$SO_2(g) + \tfrac{1}{2}O_2(g) \rightleftharpoons SO_3(g)$$

The graph illustrates how the yield of SO_3 varies with temperature.

Deduce whether the forward reaction is exothermic or endothermic.
Give a reason for your answer.

....

....

.... **(2 marks)**

Exam skills 7

1 (a) Nitrogen is oxidised by oxygen at high temperatures to form nitrogen monoxide:

$$N_2(g) + O_2(g) \rightleftharpoons 2NO(g) \qquad \Delta H^\ominus = +180 \, \text{kJ mol}^{-1}$$

   (i) Write an expression for the equilibrium constant, K_c, for this reaction.

   .... **(1 mark)**

   (ii) Explain the effect on the equilibrium yield of NO(g) if the temperature is increased.

   ....

   .... **(2 marks)**

   (iii) Explain the effect on the equilibrium yield of NO(g) if the pressure is increased.

   ....

   .... **(2 marks)**

   (iv) Explain the effect on the equilibrium yield of NO(g) if a catalyst is present.

   ....

   .... **(2 marks)**

(b) Iron(III) ions and thiocyanate ions react to form a compound ion:

$$Fe^{3+}(aq) + SCN^-(aq) \rightleftharpoons FeSCN^{2+}(aq)$$

   Explain the effect on the equilibrium position of adding ammonium chloride, which reduces the concentration of iron(III) ions.

   ....

   .... **(2 marks)**

(c) Ammonia is manufactured by the Haber process: $N_2(g) + 3H_2(g) \rightleftharpoons 2NH_3(g)$
Under the conditions chosen, usually 450 °C at 250 atmospheres pressure, the equilibrium yield is about 33% but the actual yield is about 15%.

   (i) Explain why the actual yield differs from the equilibrium yield.

   ....

   .... **(2 marks)**

   (ii) Explain why the pressure chosen may be described as a compromise pressure.

   ....

   ....

   .... **(3 marks)**

Partial pressures and K_p

1 Ammonia and hydrogen chloride react together to produce ammonium chloride:

$$NH_3(g) + HCl(g) \rightleftharpoons NH_4Cl(s)$$

Which of the following is the correct expression for K_p?

☐ **A** $K_p = \dfrac{1}{pNH_3 \times pHCl}$

☐ **B** $K_p = \dfrac{pNH_3 \times pHCl}{pNH_4Cl}$

☐ **C** $K_p = pNH_3 \times pHCl$

☐ **D** $K_p = pNH_4Cl$ **(1 mark)**

 2 State what is meant by the term partial pressure.

It is the pressure a gas would exert ....

.... **(1 mark)**

3 Nitrogen and hydrogen react together to form ammonia:

$$N_2(g) + 3H_2(g) \rightleftharpoons 2NH_3(g)$$

Initially, 3.2 mol of nitrogen and 1.8 mol of hydrogen were sealed together in a flask at 2.5 atm.

(a) Calculate the mole fraction of each gas.

N_2 ....

H_2 .... **(2 marks)**

(b) Calculate the partial pressure of each gas.

pN_2 ....

pH_2 .... **(1 mark)**

(c) At equilibrium there was 0.8 mol of NH_3.
Calculate the amounts of the other two gases present at equilibrium.

0.8 mol NH_3 forms from: N_2 ....

H_2 ....

At equilibrium: N_2 ....

H_2 .... **(2 marks)**

4 Write expressions for K_p for the following equilibria.

(a) $2NO_2(g) \rightleftharpoons N_2O_4(g)$ .... **(1 mark)**

(b) $C(s) + H_2O(g) \rightleftharpoons CO(g) + H_2(g)$ .... **(1 mark)**

(c) $SO_2(g) + \frac{1}{2}O_2(g) \rightleftharpoons SO_3(g)$ .... **(1 mark)**

Calculating K_c and K_p values

1   2.5 mol of methanoic acid is added to 3.0 mol of methanol, producing an equilibrium mixture that contains 1.0 mol of methanoic acid:

$$HCOOH(l) + CH_3OH(l) \rightleftharpoons HCOOCH_3(l) + H_2O(l)$$

Which of the following correctly shows all four equilibrium amounts?

	HCOOH /mol	CH₃OH /mol	HCOOCH₃ /mol	H₂O /mol
☐ A	1.0	0.5	2.5	2.5
☐ B	1.0	2.0	1.0	1.0
☐ C	1.0	1.5	1.5	0
☐ D	1.0	1.5	1.5	1.5

(1 mark)

2   Hydrogen can be made from steam and carbon monoxide:

$$H_2O(g) + CO(g) \rightleftharpoons H_2(g) + CO_2(g)$$

Write the expression for K_p. Use the data below to calculate its value, stating its units if any.

Gas	H₂O(g)	CO(g)	H₂(g)	CO₂(g)
Equilibrium partial pressure /atm	0.2	1.4	3.7	0.6

Expression for K_p .... **(1 mark)**

Value for K_p .... **(2 marks)**

 Maths skills

3   The table shows expressions for K_c and K_p for three different reactions. Write the units for K_c and K_p for each one, when concentrations are measured in mol dm⁻³ and partial pressures in atm.

Expression for K_c	$K_c = \dfrac{PCl_3[Cl_2]}{[PCl_5]}$	$K_c = \dfrac{[SO_3]^2}{[SO_2]^2[O_2]}$	$K_c = [NH_3][HCl]$
Units for K_c			
Expression for K_p	$K_p = \dfrac{(p_{PCl_3})(p_{Cl_2})}{(p_{PCl_5})}$	$K_p = \dfrac{(p_{SO_3})^2}{(p_{SO_2})^2(p_{O_2})}$	$K_p = (p_{NH_3})(p_{HCl})$
Units for K_p			

(3 marks)

(3 marks)

4   0.120 mol of N_2O_4 was heated in a sealed 3.00 dm³ flask.
An equilibrium was established, which contained 0.045 mol of N_2O_4:

$$N_2O_4(g) \rightleftharpoons 2NO_2(g)$$

(a) Calculate the amount of NO_2 in the equilibrium mixture.

.... **(1 mark)**

(b) Calculate the concentrations of N_2O_4 and NO_2 in the equilibrium mixture.

[N₂O₄] ....   [NO₂] .... **(1 mark)**

(c) Write the expression for K_c and calculate its value. State its units, if any.

....

....

.... **(4 marks)**

Changing K_c and K_p

1 Sulfur dioxide reacts with oxygen to form sulfur trioxide:

$$2SO_2(g) + O_2(g) \rightleftharpoons 2SO_3(g) \qquad \Delta H = -96 \text{ kJ mol}^{-1}$$

What effect would increasing the pressure have on this system?

☐ **A** The rate would decrease.

☐ **B** The equilibrium position would move to the right.

☐ **C** The equilibrium position would move to the left.

☐ **D** The equilibrium position would be unchanged.   **(2 marks)**

Maths skills

2 Calcium carbonate decomposes to form calcium oxide and carbon dioxide:

$$CaCO_3(s) \rightleftharpoons CaO(s) + CO_2(g) \qquad \Delta H = +178 \text{ kJ mol}^{-1}$$

Which of the following would increase the value of the equilibrium constant, K_p, for this equilibrium?

☐ **A** increasing the pressure

☐ **B** decreasing the pressure

☐ **C** increasing the temperature

☐ **D** decreasing the temperature   **(1 mark)**

---

3 In the Haber process, nitrogen reacts with hydrogen to produce ammonia:

$$N_2(g) + 3H_2(g) \rightleftharpoons 2NH_3(g) \qquad \Delta H = -92 \text{ kJ mol}^{-1}$$

(a) Write an expression for the equilibrium constant, K_p, for this reaction.

.... **(1 mark)**

(b) Iron is used as a catalyst for this reaction. Explain why the presence of iron does not change the value of the equilibrium constant.

....

.... **(2 marks)**

(c) Explain the effect on the equilibrium position of using osmium instead of iron as the catalyst.

....

.... **(2 marks)**

---

4 Silver carbonate decomposes to form silver oxide and carbon dioxide:

$$Ag_2CO_3(s) \rightleftharpoons Ag_2O(s) + CO_2(g)$$

An increase in temperature causes an increase in the equilibrium yield of silver oxide.

(a) State whether the forward reaction is exothermic or endothermic.

.... **(1 mark)**

(b) Explain the effect on the value of K_p of increasing the temperature.

....

.... **(2 marks)**

Had a go ☐ Nearly there ☐ Nailed it! ☐

Acids, bases and pH

1 Define the terms **Brønsted–Lowry acid** and **Brønsted–Lowry base**.

Acid: .... **(1 mark)**

Base: .... **(1 mark)**

2 Ammonia reacts with water:

$$NH_3 + H_2O \rightleftharpoons NH_4^+ + OH^-$$

Which are the Brønsted–Lowry acids?

☐ **A** NH_3 and OH^-

☐ **B** NH_3 and NH_4^+

☐ **C** H_2O and NH_4^+

☐ **D** H_2O and OH^-

Which species can donate protons?

(1 mark)

3 Explain why hydrochloric acid may be described as **monobasic**, but sulfuric acid as **dibasic**.

.... **(2 marks)**

4 State the formula of the conjugate bases of these acids:

(a) HCN: .... **(1 mark)**

(b) HF: .... **(1 mark)**

5 Identify the conjugate acid–base pairs in this reaction:

$$HClO + BrO^- \rightleftharpoons ClO^- + HBrO$$

....

.... **(2 marks)**

Maths skills

6 (a) Define **pH**.

.... **(1 mark)**

(b) The concentration of hydrogen ions in a solution is 0.255 mol dm^{-3}.
Calculate the pH of this solution, giving your answer to two decimal places.
Use the **log** or **lg** button on your calculator.

....

.... **(1 mark)**

(c) Calculate the hydrogen ion concentration of a solution that has a pH of 2.80, giving your answer to three significant figures.
Use the **10^x** button on your calculator.

....

.... **(1 mark)**

pH of acids

1 Describe the difference, in terms of degree of dissociation, between a **strong** acid and a **weak** acid.

A strong acid is ....

but a weak acid is .... **(1 mark)**

2 Calculate the pH of $0.0200 \, mol \, dm^{-3}$ nitric acid, HNO_3. Give your answer to two decimal places.

> This is a strong monobasic acid.

.... **(1 mark)**

Maths skills

3 Methanoic acid, HCOOH, is found in ant stings. Its acid dissociation constant, K_a, is $1.60 \times 10^{-4} \, mol \, dm^{-3}$ at $25 \, °C$.

(a) Write an expression for K_a of methanoic acid.

....

.... **(1 mark)**

(b) (i) Calculate the pH of a $0.275 \, mol \, dm^{-3}$ solution of methanoic acid at $25 \, °C$. Give your answer to two decimal places.

....

....

.... **(3 marks)**

(ii) State **two** assumptions made in your answer to part (i).

....

.... **(2 marks)**

4 Which of the following solutions will have the highest pH?

☐ **A** $0.100 \, mol \, dm^{-3}$ ethanoic acid

☐ **B** $0.010 \, mol \, dm^{-3}$ ethanoic acid

☐ **C** $0.010 \, mol \, dm^{-3}$ hydrochloric acid

☐ **D** $0.100 \, mol \, dm^{-3}$ hydrochloric acid **(1 mark)**

5 (a) Calculate the pK_a of chloroethanoic acid, where $K_a = 1.30 \times 10^{-3} \, mol \, dm^{-3}$ at $25 \, °C$.

.... **(1 mark)**

(b) Calculate the value for K_a of bromoethanoic acid, where p$K_a = 2.90$ at $25 \, °C$.

.... **(1 mark)**

pH of bases

1 (a) Write an equation for the dissociation of pure water.

.... **(1 mark)**

(b) Write an expression for the ionic product of water, K_w.

.... **(1 mark)**

(c) $K_w = 6.81 \times 10^{-15} \, mol^2 \, dm^{-6}$ at 20 °C.
Calculate the pH of pure water at 20 °C, giving your answer to two decimal places.

> In pure water,
> $[H^+(aq)] = [OH^-(aq)]$

....

.... **(2 marks)**

(d) $K_w = 1.47 \times 10^{-14} \, mol^2 \, dm^{-6}$ at 30 °C. Calculate the value for pK_w at 30 °C.

.... **(1 mark)**

2 Potassium hydroxide, KOH, is a strong base.

(a) State why potassium hydroxide is a strong base.

.... **(1 mark)**

> **Guided**

(b) Calculate the pH of $0.0125 \, mol \, dm^{-3}$ KOH(aq) at 25 °C, giving your answer to three significant figures.
($K_w = 1.00 \times 10^{-15} \, mol^2 \, dm^{-6}$ at 25 °C)

> Rearrange the expression for K_w.

$[OH^-(aq)] = $ ....

$[H^+(aq)] = $ ....

pH = ... **(3 marks)**

3 The table shows values for standard enthalpy changes of neutralisation, $\Delta_{neut}H^\circ$, for the reactions between sodium hydroxide solution and three acids.

Acid	$\Delta_{neut}H^\circ$ /kJ mol^{-1}
HCl(aq)	−57.9
HNO$_3$(aq)	−57.6
HCN(aq)	−11.2

(a) State why the values of $\Delta_{neut}H^\circ$ are very similar for hydrochloric acid and nitric acid.

.... **(1 mark)**

(b) Explain why the value of $\Delta_{neut}H^\circ$ for hydrogen cyanide is so different from the other two values.

....

.... **(2 marks)**

Buffer solutions

1 Define the term **buffer solution**.

....

.... **(2 marks)**

2 A buffer solution is made by mixing dilute methanoic acid, HCOOH(aq), with sodium methanoate solution, HCOONa(aq).

   (a) Write an equation for the equilibrium that forms.

.... **(1 mark)**

> **Guided**

   (b) Explain how this mixture is effective as a buffer when a small amount of dilute alkali is added.

> You need to include methanoic acid, and methanoate and hydrogen ions.

   Buffer contains ....

....

.... **(3 marks)**

3 Which of the following solutions, when mixed, could make an alkaline buffer?

   ☐ **A** methanoic acid and sodium methanoate

   ☐ **B** methanoic acid and sodium hydroxide

   ☐ **C** ammonium chloride and ammonium methanoate

   ☐ **D** ammonia and ammonium chloride   **(1 mark)**

4 The pH of blood is controlled by carbonic acid, H_2CO_3(aq), and its conjugate base. Give the formula of this conjugate base.

.... **(1 mark)**

5 A buffer solution is made by mixing $150\,cm^3$ of $0.100\,mol\,dm^{-3}$ ethanoic acid, CH_3COOH(aq), with $50\,cm^3$ of $0.200\,mol\,dm^{-3}$ sodium ethanoate solution, CH_3COONa(aq).

   (a) Calculate the concentrations of ethanoic acid and sodium ethanoate in the mixture.

> The total volume of the mixture is $200\,cm^3$ ($150\,cm^3 + 50\,cm^3$).

.... **(1 mark)**

   (b) Calculate the concentration of H^+(aq) ions in the mixture. ($K_a = 1.74 \times 10^{-5}\,mol\,dm^{-3}$ at 25 °C).

> You need your answers to part (a) and the value for K_a of ethanoic acid (given).

....

.... **(2 marks)**

   (c) Use your answer to part (b) to calculate the pH of the buffer solution at 25 °C.

.... **(1 mark)**

More pH calculations

**Maths
skills**

1 A buffer solution can be made by dissolving powdered sodium ethanoate, CH_3COONa, in $0.500 \, mol \, dm^{-3}$ ethanoic acid, $CH_3COOH(aq)$.
A school laboratory technician needs to make $200 \, cm^3$ of this buffer, with a pH of 4.92 (K_a of ethanoic acid $= 1.74 \times 10^{-5} \, mol \, dm^{-3}$ at $25 \,°C$).

(a) Calculate the hydrogen ion concentration, $[H^+(aq)]$ in this buffer at $25 \,°C$, giving your answer to three significant figures.

> You are given the pH required.

.... **(1 mark)**

(b) Calculate the concentration of sodium ethanoate required.

> You are given the concentration of ethanoic acid and its K_a, and you have calculated $[H^+(aq)]$.

....

.... **(2 marks)**

(c) Use your answer to part (b) to calculate the amount, in mol, of sodium ethanoate required.

> You are given the volume of the buffer.

.... **(1 mark)**

(d) Calculate the mass of sodium ethanoate needed, giving your answer to three significant figures.

> You will need to calculate the molar mass of CH_3COONa.

....

.... **(2 marks)**

2 The table shows the pH of nitric acid and of ethanoic acid at three different concentrations.

Acid	pH of		
	$0.25 \, mol \, dm^{-3}$ acid	$0.025 \, mol \, dm^{-3}$ acid	$0.0025 \, mol \, dm^{-3}$ acid
nitric acid	0.6	1.6	2.6
ethanoic acid	2.7	3.2	3.7

(a) Describe the change in pH for each acid when it is diluted by a factor of 10.

> Your answers should be quantitative, not qualitative.

Nitric acid ....

Ethanoic acid .... **(1 mark)**

(b) Explain why ethanoic acid behaves differently from nitric acid when it is diluted.

....

....

.... **(2 marks)**

Titration curves

1 These titration curves were obtained using $0.100 \, mol \, dm^{-3}$ solutions of different acids and bases.

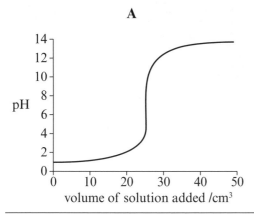

A

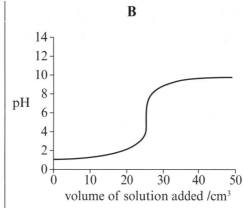

B

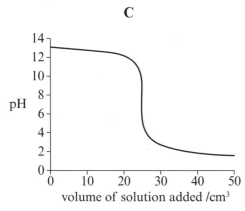

C

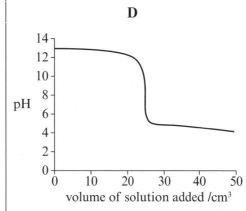

D

Identify the curve that would be produced by adding:

(a) Sodium hydroxide solution to $25 \, cm^3$ of hydrochloric acid

Curve .... **(1 mark)**

(b) Ethanoic acid solution to $25 \, cm^3$ of sodium hydroxide solution

Curve .... **(1 mark)**

(c) Ammonia solution to $25 \, cm^3$ of hydrochloric acid

Curve .... **(1 mark)**

2 On the empty graph below, sketch the titration curve that would be obtained by adding $50 \, cm^3$ of $1.00 \, mol \, dm^{-3}$ hydrochloric acid from a burette to $25 \, cm^3$ of $1.00 \, mol \, dm^{-3}$ ammonia solution.

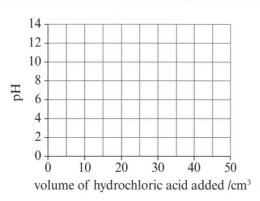

(4 marks)

Determining K_a

1   The table shows the pH ranges of some indicators.

Indicator	pK_{In}	*acidic*	pH range	*alkaline*
methyl orange	3.7	red	3.2 – 4.4	yellow
bromothymol blue	7.0	yellow	6.0 – 7.6	blue
phenolphthalein	9.3	colourless	8.2 – 10.0	red

(a)   Explain why methyl orange is a suitable indicator in the titration of a strong acid against a weak base, but phenolphthalein is not.

> What is the pH range at the equivalence point?

....

.... **(2 marks)**

(b)   State why none of the indicators is suitable for a weak acid–weak base titration.

....

.... **(1 mark)**

(c)   Explain why bromothymol blue turns blue in alkaline solutions.

> Your answer should include how the indicator forms an anion in these solutions.

....

.... **(2 marks)**

Practical skills

2   A titration was carried out to determine the concentration of some dilute ethanoic acid. In the titration, $0.100 \, \text{mol dm}^{-3}$ sodium hydroxide solution was added to $25 \, \text{cm}^3$ samples of the acid in a conical flask. The mean titre was $22.60 \, \text{cm}^3$. K_a for ethanoic acid is $1.74 \times 10^{-5} \, \text{mol dm}^{-3}$ at $25 \, ^\circ\text{C}$.

(a)   Suggest the most suitable indicator for this titration from the table in Question 1.

.... **(1 mark)**

(b)   Calculate the concentration of the ethanoic acid, giving your answer to 3 significant figures.

....

....

.... **(3 marks)**

(c)   Explain the value of the pH of the mixture in the flask when $11.30 \, \text{cm}^3$ of alkali has been added.

> This is the volume at half-equivalence ($22.60 \, \text{cm}^3 \div 2$).

....

.... **(2 marks)**

(d)   Explain why the pH of the mixture in the flask increases gradually until just before the end-point.

....

.... **(2 marks)**

Exam skills 8

1 Benzoic acid, C_6H_5COOH, is a weak acid. A 250 cm³ aqueous solution was prepared by dissolving 0.36 g of benzoic acid in water. The pH of this solution was measured using a calibrated pH meter.

(a) Calculate the molar mass of benzoic acid.
  (A_r of H = 1.0, A_r of C = 12.0, A_r of O = 16.0)

.... **(1 mark)**

(b) (i) Calculate the concentration of benzoic acid in the solution, giving your answer to three significant figures.

> You will first need to calculate the amount, in mol, of benzoic acid used to make the solution.

....

....

.... **(2 marks)**

  (ii) State an assumption you have made in your calculation in part (i).

.... **(1 mark)**

(c) The measured pH of the solution was 3.06. Calculate the concentration of H^+ ions in the solution, giving your answer to three significant figures.

.... **(1 mark)**

(d) Write an expression for the acid dissociation constant, K_a, for benzoic acid.

.... **(1 mark)**

(e) (i) Use your answers to parts (b), (c) and (d) to calculate the value for K_a.

....

.... **(2 marks)**

  (ii) State an assumption you have made in your calculation in part (i).

.... **(1 mark)**

  (iii) The accepted value for K_a at 25 °C is 6.32×10^{-5} mol dm⁻³. Apart from the assumptions made, suggest a reason for the difference between this value and the calculated value.

....

.... **(1 mark)**

Born–Haber cycles 1

1 **Standard lattice energy**, $\Delta_{lat}H^\ominus$, can be defined as the energy change when one mole of an ionic solid is formed from its gaseous ions at 100 kPa and a stated temperature, usually 298 K. Write a balanced equation, including state symbols, for the change represented by the lattice energy of magnesium chloride, $MgCl_2$.

.... **(1 mark)**

2 (a) (i) Define the term, **enthalpy change of atomisation**, $\Delta_{at}H$.

> Your answer should include the state of the atoms formed.

....

.... **(2 marks)**

(ii) Write an equation for the change represented by the enthalpy change of atomisation of bromine.

> This particular equation must include the correct state symbols.

.... **(1 mark)**

(b) (i) Define the term, **electron affinity**, Δ_{aff}.

> The change referred to involves the formation of ions from atoms.

....

.... **(2 marks)**

(ii) Write an equation for the change represented by the second electron affinity of sulfur.

.... **(1 mark)**

3 The diagram shows a Born–Haber cycle for the formation of lithium iodide.

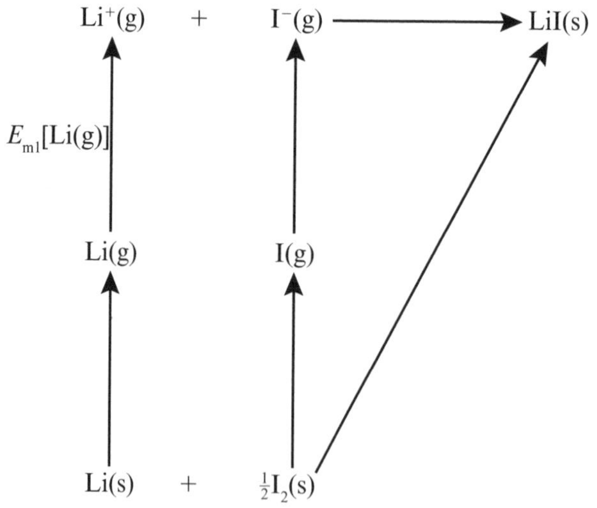

$Li^+(g)$ + $I^-(g)$ ⟶ $LiI(s)$

$E_{m1}[Li(g)]$

$Li(g)$ $I(g)$

$Li(s)$ + $\frac{1}{2}I_2(s)$

(a) Name the energy change represented by the symbol $E_{m1}[Li(g)]$.

.... **(1 mark)**

(b) Complete the diagram by adding these missing symbols:

$\Delta_{lat}H[LiI(s)]$ $\Delta_{at}H[Li(s)]$ $\Delta_{at}H[\frac{1}{2}I_2(s)]$

$\Delta_{aff1}[I(g)]$ $\Delta_f H[LiI(s)]$ **(5 marks)**

Born–Haber cycles 2

1   The table shows some data relevant to the Born–Haber cycle for the formation of magnesium oxide.

Energy change	ΔH /kJ mol^{-1}
Enthalpy change of atomisation of magnesium	+148
Enthalpy change of atomisation of oxygen	+249
First ionisation energy of magnesium	+738
Second ionisation energy of magnesium	+1451
First electron affinity of oxygen	−141
Second electron affinity of oxygen	+798
Lattice energy of magnesium oxide	−3791

(a)   Draw a Born–Haber cycle for the formation of magnesium oxide, MgO(s), showing:

     (i)   the species and their state symbols at each stage   **(3 marks)**

     (ii) the values for the energy changes for each change.   **(1 mark)**

(b)   Calculate the enthalpy of formation of magnesium oxide using these data.

....

....

....   **(2 marks)**

(c)   Explain why the second electron affinity of oxygen has a positive value.

                                 | Think about the charge on the particles involved when forming O^{2-} ions. |

....

....   **(2 marks)**

An ionic model

1 Explain which of the following ions has the greatest ability to polarise an anion: Na^+, K^+, Ca^{2+} or Ba^{2+}.

....

.... **(2 marks)**

2 Which of the following compounds has the greatest ionic character?

☐ **A** NaF

☐ **B** NaCl

☐ **C** NaBr

☐ **D** NaI

> The Data Booklet has a table of Pauling electronegativities and a table to show % ionic character.

(1 mark)

Guided **3** The table shows the lattice enthalpies of some ionic compounds, given in $kJ\,mol^{-1}$.

	Cl^-	Br^-	I^-
Li^+	848	803	759
Na^+	780	742	705
K^+	711	679	651

(a) Explain the trend in the lattice enthalpies from sodium chloride to sodium iodide.

Going down Group 7, the ionic radius ....

....

.... **(3 marks)**

(b) Explain the trend in the lattice enthalpies from lithium bromide to potassium bromide.

....

.... **(3 marks)**

4 Lattice enthalpy can be calculated using an ionic model involving electrostatic theory.

(a) Describe two assumptions of this model.

....

.... **(2 marks)**

(b) The experimentally determined lattice enthalpy for silver iodide, AgI, is $-889\,kJ\,mol^{-1}$, but its theoretically determined lattice energy is $-778\,kJ\,mol^{-1}$. Explain why the values differ so much.

....

.... **(2 marks)**

Dissolving

1 Define the term **enthalpy change of solution**, $\Delta_{sol}H$.

....

.... **(2 marks)**

2 Which of the reactions shown below has an enthalpy change equal to the enthalpy change of hydration, $\Delta_{hyd}H$, of the potassium ion?

☐ **A** $K^+(s) + \text{excess } H_2O(l) \rightarrow K^+(aq)$

☐ **B** $K^+(g) + \text{excess } H_2O(l) \rightarrow K^+(aq)$

☐ **C** $K^+(g) + 1 \text{ mol } H_2O(l) \rightarrow K^+(aq)$

☐ **D** $K^+(s) + 1 \text{ mol } H_2O(l) \rightarrow K^+(aq)$ **(1 mark)**

3 Explain why values for $\Delta_{hyd}H$ are always negative.

....

.... **(2 marks)**

4 The table shows some data about magnesium chloride, $MgCl_2$, and its ions.

	Energy /kJ mol⁻¹
Lattice energy of magnesium chloride	−2493
Enthalpy change of hydration of magnesium ions	−1920
Enthalpy change of hydration of chloride ions	−363

(a) Complete the energy cycle diagram below to represent the energy changes that occur when magnesium chloride dissolves in water. Include the lattice energy, $\Delta_{lat}H$.

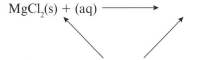

(2 marks)

(b) Use your answer to part (a) and the data provided to calculate the enthalpy change of solution of magnesium chloride.

....

.... **(2 marks)**

5 The enthalpy change of hydration for magnesium ions, Mg^{2+}, is more exothermic than the enthalpy change of hydration of potassium ions, K^+. Explain this observation.

....

.... **(2 marks)**

Entropy

1   State the natural direction of entropy change.

.... **(1 mark)**

2   Which of these substances in the solid state is likely to have the greatest standard entropy?

☐ **A** PbF_2

☐ **B** $PbCl_2$

☐ **C** $PbBr_2$

> Which molecules are the most complex?

☐ **D** PbI_2 **(1 mark)**

3   Sodium chloride crystals dissolve in water to form sodium chloride solution. Explain the change in the entropy of the system.

> State whether the entropy increases or decreases, then explain this in terms of the ways in which the particles can be arranged.

....

.... **(2 marks)**

4   The table shows some data about water in its three different states.

Substance	Entropy /$J\,K^{-1}\,mol^{-1}$
$H_2O(s)$	62.1
$H_2O(l)$	69.9
$H_2O(g)$	188.7

(a)   Give the state in which water molecules are most disordered.

.... **(1 mark)**

(b)   Describe the arrangement and movement of water molecules in liquid water.

....

.... **(2 marks)**

(c)   Explain the difference between the entropy change when water melts and when it boils.

....

....

.... **(3 marks)**

5   Which of these reactions leads to a decrease in the entropy of the system?

☐ **A** $CaCO_3(s) \rightarrow CaO(s) + CO_2(g)$

☐ **B** $N_2O_4(g) \rightarrow 2NO_2(g)$

☐ **C** $N_2(g) + 3H_2(g) \rightarrow 2NH_3(g)$

☐ **D** $NH_4Cl(s) \rightarrow NH_3(g) + HCl(g)$ **(1 mark)**

Calculating entropy changes

 Maths skills

1 Hydrogen is produced by reacting methane with steam:

$$CH_4(g) + H_2O(g) \rightarrow CO(g) + 3H_2(g)$$

The table shows some data about these reactants and products.

Substance	$\Delta_f H^\ominus$ /kJ mol^{-1}	$S^\ominus$ /J K^{-1} mol^{-1}
$CH_4(g)$	−74.8	186.2
$H_2O(g)$	−241.8	188.7
$CO(g)$	−110.5	197.6
$H_2(g)$		130.6

(a) Suggest why the standard enthalpy change of formation for hydrogen is not given in the table.

....

.... **(2 marks)**

(b) Calculate the entropy change in the system of the reaction between methane and steam.

$\Delta S^\ominus_{system} = \sum S^\ominus_{products} - \sum S^\ominus_{reactants}$

....

.... **(2 marks)**

(c) Calculate the enthalpy change of reaction, $\Delta_r H^\ominus$.

....

.... **(2 marks)**

(d) Use your answer to part (c) to calculate the entropy change in the surroundings at 298 K.

$\Delta S^\ominus_{surroundings} = \dfrac{-\Delta H^\ominus}{T}$

Remember that $\Delta H^\ominus$ is measured in kJ mol^{-1} but $\Delta S^\ominus$ is measured in J K^{-1} mol^{-1}, so you have a factor of 1000 to take into account.

....

.... **(2 marks)**

(e) Use your answers to parts (b) and (d) to calculate the total entropy change, ΔS_{total}

.... **(1 mark)**

(f) Explain the meaning of the term feasible in the context of chemical reactions.

.... **(2 marks)**

(g) Predict the temperature, to three significant figures, at which the reaction becomes feasible.

$\Delta G^\ominus = \Delta H^\ominus - T\Delta S^\ominus_{system}$

....

....

.... **(3 marks)**

Gibbs energy and equilibrium

1 Lithium carbonate decomposes at high temperature to form lithium oxide and carbon dioxide:

$$Li_2CO_3(s) \rightarrow Li_2O(s) + CO_2(g)$$

Lithium carbonate is thermodynamically stable at room temperature because:

☐ **A** the enthalpy change, ΔH, for the reaction is positive

☐ **B** the Gibbs energy, ΔG, is negative

☐ **C** the entropy change of the system, ΔS_{system}, is positive

☐ **D** the activation energy for this reaction is high.

> Thermodynamically stable is not the same as kinetically stable.

(1 mark)

[Maths skills]

2 Hydrogen reacts with iodine to produce hydrogen iodide:

$$H_2(g) + I_2(g) \rightleftharpoons 2HI(g)$$

$\Delta H^\ominus = -9.6\,kJ\,mol^{-1}$ $\Delta S^\ominus_{system} = +21.8\,J\,K^{-1}\,mol^{-1}$

(a) Calculate the Gibbs energy at 298 K, giving your answer to three significant figures.

> $\Delta G^\ominus = \Delta H^\ominus - T\Delta S^\ominus_{system}$

Remember that $\Delta H^\ominus$ and $\Delta G^\ominus$ are measured in $kJ\,mol^{-1}$ but $\Delta S^\ominus$ is measured in $J\,K^{-1}\,mol^{-1}$, so you have a factor of 1000 to take into account.

....

.... **(2 marks)**

(b) Explain whether this reaction is thermodynamically feasible at 298 K.

....

.... **(2 marks)**

(c) (i) Use the expression $\Delta G = -RT\ln K$ to calculate the value for $\ln K$ at 298 K.

> You will need to convert your value for ΔG to $J\,mol^{-1}$ first.

....

.... **(2 marks)**

(ii) Use your answer to part (i) to calculate the value for the equilibrium constant, K, at 298 K.

You will need to use the button marked e^x on the your calculator.

.... **(1 mark)**

(d) (i) Comment on the value for the equilibrium constant, K, calculated in part (c).

> In your answer, consider how small or large the value for K is, and how this affects the position of equilibrium and the equilibrium concentration of hydrogen iodide.

....

.... **(2 marks)**

(ii) Suggest why the rate of reaction is very low at 298 K.

.... **(1 mark)**

Redox and standard electrode potential

1 In sulfuric(VI) acid, H_2SO_4, the oxidation number of sulfur is +6.
This means that the sulfur in sulfuric acid:

☐ **A** is an ion with a charge of +6

☐ **B** forms a total of six covalent bonds

☐ **C** has six electrons in its outer shell

☐ **D** would have a charge of +6 if its bonding electrons were transferred completely. **(1 mark)**

2 Cr^{3+} ions can be oxidised to CrO_4^{2-} ions in alkaline solution.

(a) State, in terms of electron transfer, what is meant by the term **oxidation**.

.... **(1 mark)**

(b) Explain, in terms of oxidation number, why Cr^{3+} ions are oxidised in the reaction.

....

.... **(2 marks)**

3 State the conditions referred to by the term
standard electrode potential, $E^{\ominus}$.

> Include temperature, pressure and concentration.

....

....

.... **(3 marks)**

4 Sketch a labelled diagram to show the essential features of the standard hydrogen electrode.

(4 marks)

> **Guided**

5 (a) Explain the value of $E^{\ominus}$ for the standard hydrogen electrode.

The value is ....

because .... **(2 marks)**

(b) Explain why a reference electrode is necessary for measuring electrode potentials of half-cells.

....

.... **(2 marks)**

Measuring standard emf

1   An electrochemical cell consists of a standard hydrogen electrode placed on the left of a $Zn^{2+}(aq)|Zn(s)$ half-cell, which includes a zinc strip in zinc sulfate solution. Which one of the following does not affect the emf of the cell?

    ☐ **A**   the size of the zinc strip

    ☐ **B**   the temperature of the zinc sulfate solution

    ☐ **C**   the concentration of the zinc sulfate solution

    ☐ **D**   the pressure of the hydrogen gas in the standard electrode   **(1 mark)**

Practical skills

2   Describe how the following half-cells could be set up in glass beakers.

    (a)   An $Mg^{2+}(aq)|Mg(s)$ half-cell.

    ....   **(1 mark)**

    (b)   A half-cell with which the electrode potential for this reaction could be measured:

       $Fe^{2+} \rightleftharpoons Fe^{3+} + e^-$

> What substance would you use for the electrode?

    ....

    ....   **(2 marks)**

3   The diagram shows an electrochemical cell set up under standard conditions.

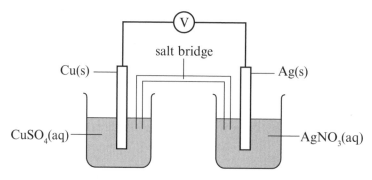

    (a)   Give a suitable salt for use in the salt bridge.

    ....   **(1 mark)**

    (b)   Describe the function of the salt bridge.

    ....   **(1 mark)**

    (c)   Give the cell diagram for the cell using the conventional representation of half-cells.

> For example, the standard hydrogen electrode is given as $Pt(s)|H_2(g)|H^+(aq)$.

    ....   **(2 marks)**

    (d)   The table shows the standard electrode potentials for the two half-cells. Calculate the standard emf, $E^{\ominus}_{cell}$, of the cell.

Electrode system	$E^{\ominus}$/V
$Ag^+ + e^- \rightleftharpoons Ag$	+0.80
$Cu^{2+} + 2e^- \rightleftharpoons Cu$	+0.34

    ....

    ....   **(1 mark)**

Predicting reactions

Use these electrode potentials to answer the questions.

Right-hand electrode system			$E^\ominus$/V
$Fe^{2+} + 2e^-$	$\rightleftharpoons$	Fe	−0.44
$\frac{1}{2}I_2 + e^-$	$\rightleftharpoons$	I^-	+0.54
$Fe^{3+} + e^-$	$\rightleftharpoons$	Fe^{2+}	+0.77
$Ag^+ + e^-$	$\rightleftharpoons$	Ag	+0.80

1　Which species in the list is the most powerful reducing agent?

　☐ **A** Fe^{2+}

　☐ **B** Fe

　☐ **C** Ag

　☐ **D** Ag^+　　　　　　　　　　　　　　　　　　　　　　　　　　　　**(1 mark)**

2　Iron filings are added to silver nitrate solution, $AgNO_3$(aq).

　(a)　Write the ionic equation for the reaction between silver ions and iron to form silver atoms and iron(II) ions. Calculate $E^\ominus_{cell}$ for this reaction.

　...

　...　**(2 marks)**

　(b)　Write the ionic equation for the reaction between silver ions and iron(II) ions to form silver atoms and iron(III) ions. Calculate $E^\ominus_{cell}$ for this reaction.

　...

　...　**(2 marks)**

　(c)　Use your answers to parts (a) and (b) to predict the final species of iron formed if excess iron filings are added to silver nitrate solution.
　Explain your answer.

　┌─────────────────────────────┐
　│ Feasible reactions under   │
　│ standard conditions have   │
　│ a positive value of $E^\ominus$. │
　└─────────────────────────────┘

　...

　...　**(2 marks)**

3　(a)　Use $E^\ominus$ values to predict whether silver should react with iodine.

　　　$E^\ominus_{cell}$ for feasible reaction = ...

　...　**(2 marks)**

　(b)　Use $E^\ominus$ values to predict whether this disproportionation reaction occurs under standard conditions:

　　　　$3Fe^{2+}$(aq) $\rightarrow$ Fe(s) $+ 2Fe^{3+}$(aq)

　...

　...　**(2 marks)**

Limitations to predictions

1 Zn reacts with copper(II) ions in solution:

$$Zn(s) + Cu^{2+}(aq) \rightarrow Zn^{2+}(aq) + Cu(s).$$

Under standard conditions, the value of the equilibrium constant, K_c, for this reaction is greater than 1. This means that, for this reaction:

☐ **A** $E^{\ominus}_{cell}$ and $\Delta S^{\ominus}_{total}$ both have negative values

☐ **B** $E^{\ominus}_{cell}$ is positive and $\Delta S^{\ominus}_{total}$ is negative

☐ **C** $E^{\ominus}_{cell}$ is negative and $\Delta S^{\ominus}_{total}$ is positive

☐ **D** $E^{\ominus}_{cell}$ and $\Delta S^{\ominus}_{total}$ both have positive values. **(1 mark)**

2 Use these standard electrode potentials in this question.

Right-hand electrode system			$E^{\ominus}$/V
$Cu^{2+} + 2e^-$	$\rightleftharpoons$	Cu	+0.34
$NO_3^- + 2H^+ + e^-$	$\rightleftharpoons$	$NO_2 + H_2O$	+0.80

(a) Combine these half-equations to show the reaction between copper and nitric acid, $HNO_3(aq)$.

.... **(1 mark)**

(b) Determine whether this reaction is feasible under standard conditions.

....

.... **(2 marks)**

(c) Suggest why the reaction proceeds more readily when concentrated nitric acid is used.

....

.... **(2 marks)**

3 Use these standard electrode potentials in this question.

Right-hand electrode system			$E^{\ominus}$/V
$\frac{1}{2}O_2 + 2H^+ + 2e^-$	$\rightleftharpoons$	H_2O	+1.23
$\frac{1}{2}Cl_2 + e^-$	$\rightleftharpoons$	Cl^-	+1.36

(a) Write an equation for the redox reaction between chlorine and water, and explain why it should occur.

....

....

.... **(2 marks)**

(b) Suggest why this reaction does not usually happen in the dark.

.... **(1 mark)**

4 Explain the effect on the E_{cell} value for this cell if the zinc ion concentration is increased:

> What effect will this change have on the equilibrium in the right-hand electrode?

$$Mg(s)|Mg^{2+}(aq) \vdots\vdots Zn^{2+}(aq)|Zn(s).$$

....

....

.... **(3 marks)**

Storage cells and fuel cells

1 Use these standard electrode potentials in this question.

Right-hand electrode system	$E^\ominus$/V
$2H_2O + 2e^- \rightleftharpoons 2OH^- + H_2$	−0.83
$2H^+ + 2e^- \rightleftharpoons H_2$	0.00
$\frac{1}{2}O_2 + H_2O + 2e^- \rightleftharpoons 2OH^-$	+0.40
$\frac{1}{2}O_2 + 2H^+ + 2e^- \rightleftharpoons H_2O$	+1.23

(a) Explain why a hydrogen–oxygen fuel cell does not need to be electrically recharged.

....

....

.... **(1 mark)**

(b) (i) Calculate $E^\ominus_{cell}$ for a hydrogen–oxygen fuel cell operating under acidic conditions.

....

.... **(1 mark)**

(ii) Write a balanced equation for the overall reaction that happens in this fuel cell.

.... **(1 mark)**

(c) (i) Calculate $E^\ominus_{cell}$ for a hydrogen–oxygen fuel cell operating under alkaline conditions.

.... **(1 mark)**

(ii) Write a balanced equation for the overall reaction that happens in this fuel cell.

....

.... **(2 marks)**

(d) Comment on your answers to parts (b) and (c).

> Are the answers the same or different, and what is the reason for this?

....

.... **(2 marks)**

2 Use these standard electrode potentials in this question.

Right-hand electrode system	$E^\ominus$/V
$Cd(OH)_2 + 2e^- \rightleftharpoons Cd + 2OH^-$	−0.81
$Ni(OH)_3 + e^- \rightleftharpoons Ni(OH)_2 + OH^-$	+0.48

(a) A nickel–cadmium rechargeable cell uses these reactions.
Calculate the $E^\ominus_{cell}$ when it discharges.

.... **(1 mark)**

(b) Suggest why the measured potential difference may be less than your answer to part (a).

....

.... **(1 mark)**

(c) Write a balanced equation for the overall reaction that happens when the cell is being charged.

.... **(2 marks)**

Exam skills 9

1 (a) Fuel cells can use fuels other than hydrogen, such as ethanol, C_2H_5OH.

    (i) Assuming that the carbon atoms and hydrogen atoms in the fuel become fully oxidised, write an equation for the overall reaction that happens in an ethanol–oxygen fuel cell.

.... **(1 mark)**

    (ii) The reaction at the oxygen electrode is the same reaction that occurs in a hydrogen–oxygen fuel cell operating under acidic conditions.
Write the half-equation.

.... **(1 mark)**

    (iii) Deduce the half-equation for the reaction at the ethanol electrode.

....

.... **(2 marks)**

    (iv) $E^\ominus = +1.23$ V for the oxygen electrode.
Calculate $E^\ominus$ for the ethanol electrode ($E^\ominus_{cell} = 1.01$ V).

.... **(1 mark)**

(b) The batteries in vehicles with petrol engines or diesel engines contain six lead–acid cells. The table shows the standard electrode potentials involved.

Right-hand electrode system	$E^\ominus$ /V
$PbSO_4 + H^+ + 2e^- \rightleftharpoons Pb + HSO_4^-$	−0.36
$PbO_2 + HSO_4^- + 3H^+ + 2e^- \rightleftharpoons PbSO_4 + 2H_2O$	+1.69

    (i) Calculate $E^\ominus_{cell}$ when a single cell is being discharged.

.... **(1 mark)**

    (ii) Write a balanced equation for the overall reaction that happens during recharging.

....

.... **(2 marks)**

    (iii) Suggest the minimum potential difference needed to recharge a lead–acid car battery, and justify your answer.

> Remember that the battery contains six cells.

....

.... **(2 marks)**

Redox titrations

1. Potassium manganate(VII) solution, acidified with dilute sulfuric acid, is added from a burette to determine the concentration of iron(II) ions by titration. At the end-point, the solution in the flask:
   - ☐ **A** becomes pink
   - ☐ **B** becomes purple
   - ☐ **C** becomes brown
   - ☐ **D** becomes colourless.   **(1 mark)**

Practical skills

2. Iron(II) sulfate is an ingredient of lawnsand, used as a lawn fertiliser and to kill moss. A 9.35 g sample of lawnsand was mixed with 100 cm³ of dilute sulfuric acid and filtered. The filtrate was made up to 250 cm³ in a volumetric flask with de-ionised water. When 25.0 cm³ portions were titrated with 5.00×10^{-3} mol dm⁻³ potassium manganate(VII), the mean titre was 27.60 cm³.

   (a) Write a balanced equation for the reaction that occurs in the titration.

   ....

   .... **(1 mark)**

   > The half-equations are: $MnO_4^- + 8H^+ + 5e^- \rightarrow Mn^{2+} + 4H_2O$ and $Fe^{2+} \rightarrow Fe^{3+} + e^-$

   (b) Calculate the amount in mol of manganate(VII) ions in the mean titre.

   ....

   .... **(1 mark)**

   (c) Use your answers to parts (a) and (b) to calculate the amount in mol of iron(II) ions in each 25 cm³ portion.

   ....

   .... **(1 mark)**

   (d) Calculate the amount in mol of iron(II) ions in the original sample of lawnsand.

   ....

   > A solution produced from the original sample was added to a volumetric flask.

   ....

   .... **(1 mark)**

   (e) Calculate the mass of iron in the original sample of lawnsand. (Molar mass of Fe = 55.8 g mol⁻¹)

   ....

   .... **(1 mark)**

   (f) Calculate the percentage by mass of iron in the lawnsand, giving your answer to three significant figures.

   > You know the mass of iron and the mass of lawnsand used.

   ....

   .... **(1 mark)**

   (g) State why sulfuric acid is added in this experiment.

   ....

   .... **(1 mark)**

   (h) Explain why hydrochloric acid should not be used instead of sulfuric acid in this experiment.

   ....

   .... **(2 marks)**

101

d-block atoms and ions

Guided   **1**   Write the full electronic configurations of the following atoms.

(a)   Scandium, $Z = 21$

$1s^2\ 2s^2\ 2p^6\ 3s^2\ 3p^6$ .... **(1 mark)**

(b)   Chromium, $Z = 24$

.... **(1 mark)**

2   Write the full electronic configurations of the following ions.

(a) Fe^{3+}, $Z = 26$

.... **(1 mark)**

(b) Mn^{4+}, $Z = 25$

.... **(1 mark)**

3   (a)   Define the term **transition metal**.

....

.... **(2 marks)**

(b)   Explain why scandium is not a transition metal.

....

.... **(2 marks)**

4   The electronic structure of titanium, Ti, is $1s^2\ 2s^2\ 2p^6\ 3s^2\ 3p^6\ 3d^2\ 4s^2$. Which of the following compounds is **not** likely to exist?

☐ **A** $TiCl_4$

☐ **B** K_2TiO_3

☐ **C** K_2TiF_6

☐ **D** K_2TiO_4

> What are the oxidation numbers of each element in the compounds?

(1 mark)

5   Complete the electron-in-boxes diagram for the electronic configuration of:

(a)   copper, $Z = 29$

                                  3d   **4s**

[Ar] ☐☐☐☐☐   ☐

(1 mark)

(b) V^{3+}, $Z = 23$

                                  3d   **4s**

[Ar] ☐☐☐☐☐   ☐

(1 mark)

Ligands and complex ions

1 Copper(II) ions can form complex ions with neutral ligands, such as NH_3, and with negatively charged ligands, such as OH^-. The bonding between the ligands and Cu^{2+} ions in these complexes is:

	NH_3	OH^-
☐ **A**	dative covalent	ionic
☐ **B**	dative covalent	dative covalent
☐ **C**	ionic	dative covalent
☐ **D**	ionic	ionic

(1 mark)

2 The tetrachlorocobaltate(II) ion, $[CoCl_4]^{2-}$, is a complex ion.

   (a) (i) Identify the metal ion and the ligand in this complex.

.... **(2 marks)**

       (ii) Define the term **ligand**.

.... **(2 marks)**

   (b) State the coordination number of this complex, and give a reason for your answer.

....

.... **(2 marks)**

3 State why H_2O may be described as a **monodentate** ligand.

.... **(1 mark)**

4 The ethandioate ion, $C_2O_4{}^{2-}$, is a bidentate ligand.
Draw its full displayed formula, identifying the features that allow it to act as a bidentate ligand.

(2 marks)

5 The diagram shows the structure of $EDTA^{4-}$. This ion can act as a **multidentate** ligand.

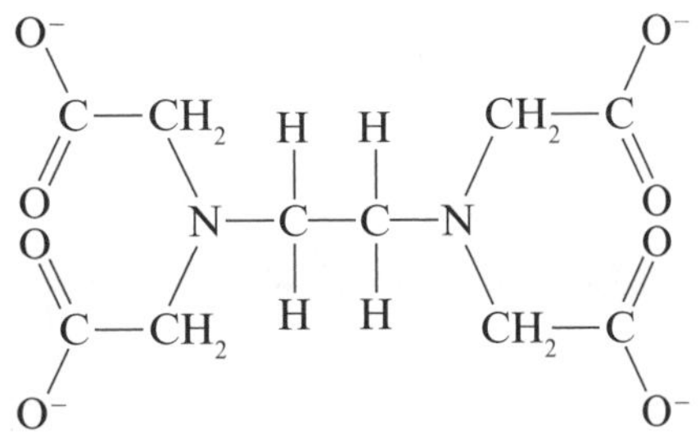

   (a) State the number of lone pairs of electrons available to form dative bonds with a metal ion.

.... **(1 mark)**

   (b) Draw the positions of these lone pairs of electrons on the diagram above.   **(1 mark)**

Shapes of complexes

1 The shapes of the complexes $[CoCl_4]^{2-}$ and $[Ag(NH_3)_2]^+$ are:

	$[CoCl_4]^{2-}$	$[Ag(NH_3)_2]^+$
☐ A	square planar	tetrahedral
☐ B	octahedral	linear
☐ C	tetrahedral	linear
☐ D	tetrahedral	V-shaped

(1 mark)

2 The anti-cancer drug *cis*-platin has the formula $[PtCl_2(NH_3)_2]$.

(a) Draw diagrams in the spaces below to show the structures of *cis*-platin and *trans*-platin.

cis-platin	*trans*-platin

(3 marks)

(b) Name the shape of *cis*-platin.

.... **(1 mark)**

(c) Explain why platin is supplied for use in cancer treatment as the *cis* isomer alone.

....

.... **(2 marks)**

3 Copper(II) ions can form the complex ion hexaaquacopper(II), $[Cu(H_2O)_6]^{2+}$.

(a) Name the shape of this complex ion, and state the bond angle.

.... **(2 marks)**

(b) In the presence of excess chloride ions, the complex ion tetrachlorocuprate(II), $[CuCl_4]^{2-}$ forms.

(i) Name the shape of this complex ion, and state the bond angle.

.... **(2 marks)**

(ii) Describe what happens to the coordination number when $[Cu(H_2O)_6]^{2+}$ forms $[CuCl_4]^{2-}$.

.... **(1 mark)**

(iii) Explain the difference in the shape of the two complex ions, $[Cu(H_2O)_6]^{2+}$ and $[CuCl_4]^{2-}$.

....

....

> Apart from charge, how do the ligands involved differ from one another?

....

.... **(2 marks)**

Colours

1 (a) Complete the electron-in-boxes diagram for the d-orbitals in copper ions, Cu^{2+}.

(1 mark)

(b) The presence of ligands such as chloride ions cause the d-orbitals in the Cu^{2+} ions in the $[CuCl_4]^{2-}$ complex ion to split. The five d-orbitals become three d-orbitals at a higher energy level, and two d-orbitals at a lower energy level.

(i) Complete the diagram for the d-orbitals in copper ions in the non-excited state.

(1 mark)

(ii) Complete the diagram for the d-orbitals in copper ions in the excited state.

(1 mark)

(iii) Explain why $[CuCl_4]^{2-}$ appears coloured in solution.

....

.... **(2 marks)**

2 The hexaaquacopper(II) ion, $[Cu(H_2O)_6]^{2+}$, is blue in solution because:

☐ **A** electrons absorb light in the blue part of the visible spectrum and remaining frequencies are seen

☐ **B** electrons are excited and emit light in the blue part of the visible spectrum as they return to the ground state

☐ **C** electrons are excited and emit light in the red part of the visible spectrum as they return to the ground state

☐ **D** electrons absorb light in the red part of the visible spectrum and remaining frequencies are seen. **(1 mark)**

3 Not all complex ions are coloured in solution.

(a) Explain why $[Sc(H_2O)_6]^{3+}$ is colourless in solution.

....

....

.... **(3 marks)**

(b) Aluminium ions, Al^{3+}, can form complex ions. Explain why $[Al(OH)_4]^-$ is colourless in solution.

.... **(1 mark)**

Colour changes

1 When a solution containing $[Cr(H_2O)_6]^{3+}$ ions is warmed in the presence of chloride ions, Cl^-, $[Cr(H_2O)_4Cl_2]^+$ ions form. The colour of the solution changes because:

☐ **A** there is a change in the ligands in the complex

☐ **B** the oxidation number of chromium ion changes

☐ **C** the coordination number changes

☐ **D** the shape of the complex ion changes. **(1 mark)**

2 Haemoglobin is an iron(II) complex containing a ligand.

(a) State whether this ligand is monodentate, bidentate or multidentate.

.... **(1 mark)**

(b) Describe how nitrogen atoms in the ligand bond to iron(II) ions.

....

.... **(2 marks)**

(c) A reaction occurs when an oxygen molecule bound to haemoglobin is replaced by a carbon monoxide molecule. Bright red oxyhaemoglobin becomes cherry red carboxyhaemoglobin.

(i) State the type of reaction involved.

.... **(1 mark)**

(ii) Explain why there is a colour change as a result of this reaction.

....

....

> Why are complexes coloured? What factors affect this process?

....

.... **(3 marks)**

3 For each of the following situations, give the formula of the complex formed, and state why a colour change occurs.

(a) A solution containing $[Fe(H_2O)_6]^{2+}$ changes from pale green to yellow brown on standing in air.

.... **(2 marks)**

(b) A solution containing $[Cu(H_2O)_6]^{2+}$ changes from pale blue to deep blue when concentrated ammonia is added.

.... **(2 marks)**

(c) A solution containing $[Co(H_2O)_6]^{2+}$ changes from pink to blue when hydrochloric acid is added.

....

.... **(3 marks)**

Vanadium chemistry

1 (a) Which row contains the correct colour of each ion in solution?

	V^{2+}	V^{3+}	VO_2^+	VO^{2+}
☐ **A**	yellow	blue	green	purple
☐ **B**	purple	green	blue	yellow
☐ **C**	purple	green	yellow	blue
☐ **D**	yellow	blue	purple	green

(1 mark)

(b) State the oxidation state of vanadium in each of the ions shown in part (a).

V^{2+} .... V^{3+} ....

VO_2^+ .... VO^{2+} ....   **(3 marks)**

You will need these standard electrode potential values to answer some of these questions.

Right-hand electrode system	$E^{\ominus}$/V
$Zn^{2+} + 2e^- \rightleftharpoons Zn$	-0.76
$V^{3+} + e^- \rightleftharpoons V^{2+}$	-0.26
$VO^{2+} + 2H^+ + e^- \rightleftharpoons V^{3+} + H_2O$	$+0.34$
$VO_2^+ + 2H^+ + e^- \rightleftharpoons VO^{2+} + H_2O$	$+1.00$

> The Data Booklet given to you in examinations contains a table of standard electrode potentials.

2 Ammonium trioxovanadate(V), NH_4VO_3, forms the dioxovanadium(V) ion, VO_2^+, in acidic conditions. This can be reduced using zinc with sulfuric acid or hydrochloric acid.

 (a) (i) Write an equation to show the reduction of vanadium(III) ions to vanadium(II) ions using zinc.

.... **(1 mark)**

    (ii) Explain, using electrode potentials, whether this reaction is feasible under standard conditions.

> Calculate $E^{\ominus}_{right} - E^{\ominus}_{left}$ where the reduction half-equation refers to $E^{\ominus}_{right}$.

....

.... **(2 marks)**

(b) (i) Write an equation to show the reduction of VO_2^+ ions to V^{2+} ions using zinc.

....

....

.... **(2 marks)**

    (ii) State why this reaction is feasible under standard conditions.

.... **(1 mark)**

Chromium chemistry

1 (a) Which row contains the correct colour of each ion in solution?

	Cr^{2+}	Cr^{3+}	$Cr_2O_7^{2-}$	CrO_4^{2-}
☐ **A**	yellow	orange	green	blue
☐ **B**	green	blue	orange	yellow
☐ **C**	blue	green	yellow	orange
☐ **D**	blue	green	orange	yellow

(1 mark)

(b) State the oxidation state of chromium in each of the ions shown in part (a).

Cr^{2+} ....

Cr^{3+} ....

CrO_4^{2-} ....

$Cr_2O_7^{2-}$ ....   **(3 marks)**

You will need these standard electrode potential values to answer some of these questions.

Right-hand electrode system	$E^{\ominus}/V$
$Zn^{2+} + 2e^- \rightleftharpoons Zn$	−0.76
$Cr^{3+} + e^- \rightleftharpoons Cr^{2+}$	−0.41
$\frac{1}{2}O_2 + 2H^+ + 2e^- \rightleftharpoons H_2O$	+1.23
$Cr_2O_7^{2-} + 14H^+ + 6e^- \rightleftharpoons 2Cr^{3+} + 7H_2O$	+1.33

2 Chromium can be reduced using zinc with sulfuric acid or hydrochloric acid.

(a) Write an equation to show the reduction of $Cr_2O_7^{2-}$ ions to chromium(III) using zinc.

....   **(1 mark)**

(b) Explain, using electrode potentials, whether this reaction is feasible under standard conditions.

....

....   **(2 marks)**

3 Cr^{2+} ions are oxidised to Cr^{3+} ions in acidic conditions when exposed to oxygen in air.

(a) Write an equation to represent this reaction.

> You will need to select an appropriate reduction half-equation from the table above, and then combine it correctly with the half-equation for the oxidation of Cr^{2+} ions.

....

....   **(2 marks)**

(b) Explain, using electrode potentials, whether this reaction is feasible under standard conditions.

....

....   **(2 marks)**

4 Write an equation to represent the conversion of $Cr_2O_7^{2-}$ to CrO_4^{2-}.

....   **(1 mark)**

Reactions with hydroxide ions

1 Copper(II) sulfate was dissolved in deionised water to form a solution containing $[Cu(H_2O)_6]^{2+}$(aq) ions. When a few drops of sodium hydroxide solution were added to the solution, a precipitate of the complex $[Cu(H_2O)_4(OH)_2]$(s) formed.

(a) State what is meant by a **precipitate**.

When two different solutions are mixed, ...

.... **(1 mark)**

(b) Write an equation, including state symbols, to represent the reaction.

.... **(2 marks)**

(c) The reaction is an acid–base reaction (not a ligand exchange reaction).

(i) Explain how hydrogen ions are released from the complex ion, $[Cu(H_2O)_6]^{2+}$.

.... **(2 marks)**

(ii) State the role of the hydroxide ions in this reaction.

.... **(1 mark)**

2 Sodium hydroxide solution is added to an aqueous solution of a transition metal compound. A grey-green precipitate forms, which dissolves when excess sodium hydroxide solution is added.
Which of the following metal ions was present in the original transition metal compound?

☐ **A** Fe^{2+}

☐ **B** Cr^{2+}

☐ **C** Cr^{3+}

☐ **D** Fe^{3+} **(1 mark)**

3 Complete the table to show the colour of each complex ion or complex.

$[Cu(H_2O)_6]^{2+}$		$[Co(H_2O)_6]^{2+}$	
$[Cu(H_2O)_4(OH)_2]$		$[Co(H_2O)_4(OH)_2]$	

(4 marks)

4 A solution containing $[Fe(H_2O)_6]^{2+}$(aq) ions forms a precipitate when sodium hydroxide solution is added. On standing, this precipitate gradually changes colour.

(a) State the colour of each precipitate.

First precipitate ....

Second precipitate .... **(2 marks)**

(b) State the type of reaction involved.

.... **(1 mark)**

Reactions with ammonia

1   Copper(II) sulfate was dissolved in de-ionised water to form a solution containing $[Cu(H_2O)_6]^{2+}$(aq) ions. When a few drops of dilute ammonia solution were added to the solution, a precipitate of the complex $[Cu(H_2O)_4(OH)_2]$(s) formed.

   (a)   (i)   Write an equation, including state symbols, to represent the reaction.

     .... **(2 marks)**

     (ii) State the type of reaction involved.

     .... **(1 mark)**

     (iii) Explain the role of ammonia in this reaction.

     ....

     ....

     .... **(3 marks)**

   (b)   The precipitate dissolves when excess ammonia solution is added.

     (i)   Give the formula of the soluble complex ion formed.

> Only four of the ligands in this particular precipitate are involved.

     .... **(1 mark)**

     (ii) State the role of ammonia in this reaction.

     .... **(1 mark)**

2   You have a mixture of iron(II) hydroxide and cobalt(II) hydroxide. Which of the following reagents would allow you to separate iron(II) from cobalt(II) by filtration?

   ☐   **A**   ammonia solution

   ☐   **B**   sodium hydroxide solution

   ☐   **C**   dilute hydrochloric acid

   ☐   **D**   dilute nitric acid   **(1 mark)**

3   (a)   Complete the table to show the colour of each complex ion or complex.

$[Co(H_2O)_4(OH)_2]$		$[Cr(H_2O)_3(OH)_3]$	
$[Co(NH_3)_6]^{2+}$		$[Cr(NH_3)_6]^{3+}$	

(4 marks)

   (b)   Write balanced equations, including state symbols, for:

     (i)   the reaction of $[Co(H_2O)_4(OH)_2]$ with excess ammonia

     .... **(2 marks)**

     (ii) the reaction of $[Cr(H_2O)_3(OH)_3]$ with excess ammonia.

     .... **(2 marks)**

Ligand exchange

1 (a) EDTA^{4-} forms a complex with cobalt(II) ions in aqueous solution:

$$[Co(H_2O)_6]^{2+}(aq) + EDTA^{4-}(aq) \rightarrow [Co(EDTA)]^{2-}(aq) + 6H_2O(l)$$

(i) Predict the change in ΔS_{system} in this reaction.

| ΔS_{system} will increase or decrease. |

.... **(1 mark)**

(ii) Explain, in terms of the particles present, your answer to part (i).

....

.... **(2 marks)**

(iii) Explain, in terms of entropy, why $[Co(EDTA)]^{2-}$ is more stable than $Co(H_2O)_6]^{2+}$.

....

.... **(2 marks)**

(b) Ammonia also forms a complex with cobalt(II) ions:

$$[Co(H_2O)_6]^{2+}(aq) + 6NH_3(aq) \rightarrow [Co(NH_3)_6]^{2+}(aq) + 6H_2O(l)$$

Suggest why there is little difference in stability between $[Co(H_2O)_6]^{2+}$ and $[Co(NH_3)_6]^{2+}$.

....

.... **(2 marks)**

2 (a) Complete the table below. In your diagrams, include the bond angles.

Complex ion	$[Co(NH_3)_6]^{2+}$	$[CoCl_4]^{2-}$
Shape		
Name of shape		
Coordination number		

(8 marks)

(b) Explain why these complex ions have different colours, even though both contain a Co^{2+} ion.

....

.... **(3 marks)**

Heterogeneous catalysis

Guided 1 Transition metals and their compounds can act as heterogeneous catalysts. State the meaning of the term heterogeneous with reference to catalysts.

A heterogeneous catalyst is in a ...... *different state to the reactants it is catalysing.* ......

(1 mark)

2 Which of these does not happen in a reaction catalysed by a heterogeneous catalyst?

☐ **A** weakening of bonds in the reactant molecules

☐ **B** adsorption of reactant molecules onto the surface of the catalyst

☒ **C** an overall change in oxidation number of the metal in the catalyst

☐ **D** desorption of product molecules from the surface of the catalyst **(1 mark)**

3 One of the stages in the manufacture of sulfuric acid by the contact process involves the reaction between sulfur dioxide and oxygen:

$$2SO_2(g) + O_2(g) \rightleftharpoons 2SO_3(g)$$

The reaction is catalysed by V_2O_5.

(a) (i) Write an equation to show the reaction between V_2O_5 and sulfur dioxide to produce SO_3.

$V_2O_5 + SO_2 \rightarrow V_2O_4 + SO_3$ **(1 mark)**

(ii) Write an equation to show the reaction between the vanadium product from part (i) and oxygen, producing V_2O_5.

$V_2O_4 + \frac{1}{2}O_2 \rightarrow V_2O_5$ **(1 mark)**

(b) Explain, in terms of oxidation numbers and your answers to part (a), why V_2O_5 acts as a catalyst in the production of sulfur trioxide from sulfur dioxide.

Because the vanadium has variable oxidation states, and so can attract oxygen and SO_2 to produce SO_3. It is also regenerated. **(3 marks)**

4 Vehicle catalytic converters decrease carbon monoxide and nitrogen monoxide emissions from internal combustion engines.

(a) Explain why it is important to decrease emissions of carbon monoxide and nitrogen monoxide.

Carbon monoxide is toxic, Nitrogen monoxide contributes to acid rain. **(2 marks)**

(b) Write an equation for the reaction between these two gases, forming carbon dioxide and nitrogen.

$CO + NO \rightarrow CO_2 + N$ **(1 mark)**

(c) Catalytic converters also catalyse the reaction of unburnt hydrocarbons with oxygen. Write an equation to show the complete oxidation of octane, C_8H_{18}.

$C_8H_{18} + \frac{25}{2}O_2 \rightarrow 8CO_2 + 9H_2O$ **(1 mark)**

Homogeneous catalysis

Guided 1 Transition metals and their compounds can act as homogeneous catalysts. State the meaning of the term **homogeneous** with reference to catalysts.

A homogeneous catalyst is in the ....

.... **(1 mark)**

2 Iron(III) ions can catalyse the reaction between peroxodisulfate ions, $S_2O_8^{2-}$, and iodide ions, I^-.

(a) Write an equation for the reaction that happens in solution between these ions.

> Sulfate ions and iodine are produced.

.... **(1 mark)**

(b) Suggest why the activation energy for this reaction is high.

.... **(1 mark)**

(c) (i) Write an equation to show the reaction between Fe^{3+} ions and I^- ions.

.... **(1 mark)**

(ii) Write an equation to show the reaction between Fe^{2+} ions and $S_2O_8^{2-}$ ions.

.... **(1 mark)**

(d) Explain, in terms of oxidation numbers and your answers to parts (b) and (c), why Fe^{3+} ions act as a catalyst in the reaction between peroxodisulfate ions and iodide ions.

....

....

.... **(4 marks)**

(e) Explain whether iron(II) ions could also catalyse the reaction.

....

.... **(2 marks)**

3 Ethanedioate ions, $C_2O_4^{2-}$, react with manganate(VII) ions, MnO_4^-, in acidic solution. The reaction is autocatalysed.

(a) State the meaning of the term **autocatalysed**.

.... **(1 mark)**

(b) State the formula of the autocatalyst in this reaction.

.... **(1 mark)**

Exam skills 10

1 (a) Give the full electronic configurations for:

   (i) an iron atom, $Z = 26$

   .... **(1 mark)**

   (ii) an Fe^{2+} ion.

   .... **(1 mark)**

   (b) Explain why iron(II) compounds are coloured.

   ....

   ....

   .... **(3 marks)**

   (c) Explain why scandium is classified as a d-block element but not as a transition element.

   ....

   .... **(2 marks)**

2 $[Cu(H_2O)_6]^{2+}$ ions react with $EDTA^{4-}$ ions to form $[Cu(EDTA)]^{2-}$ ions.

   (a) Write an equation to represent the reaction, and state the type of reaction involved.

   ....

   .... **(2 marks)**

   (b) Explain why $[Cu(EDTA)]^{2-}$ is a more thermodynamically stable complex than $[Cu(H_2O)_6]^{2+}$.

   ....

   .... **(2 marks)**

   (c) Explain why $EDTA^{4-}$ may be described as a multidentate ligand.

   .... **(2 marks)**

3 A student added aqueous sodium hydroxide solution to chromium(III) sulfate solution. Grey-green precipitate of $[Cr(H_2O)_3(OH)_3]$(s) formed. This redissolved to form a dark green solution when excess sodium hydroxide solution was added, and a green solution when excess dilute sulfuric acid was added.

   (a) Write equations for the formation of these solutions.

   With sodium hydroxide ....

   With sulfuric acid .... **(2 marks)**

   (b) Explain why chromium(III) hydroxide is amphoteric.

   ....

   .... **(2 marks)**

Measuring reaction rates

Guided

1 Describe the meaning of the term **rate of reaction**.

The rate of a reaction is the change in ....

.... **(1 mark)**

Practical skills

2 The rate of reaction involving a gaseous product can be determined by measuring the volume of gas produced, or the change in mass as the product leaves the reaction mixture.

(a) Calculate the mass of 5.0×10^{-3} mol of the following gases.

Hydrogen: ....

Carbon dioxide: .... **(2 marks)**

(b) Calculate the volume, in cm^3, of 5.0×10^{-3} mol of gas at r.t.p. Give your answer to two significant figures. (The molar volume of an ideal gas at r.t.p. is $24 \, dm^3 \, mol^{-1}$.)

.... **(1 mark)**

(c) The maximum capacity of a typical gas syringe is $100 \, cm^3$.
Use your answers to parts (a) and (b) to explain:

(i) why the production of carbon dioxide may be measured using a gas syringe or a balance.

....

.... **(2 marks)**

(ii) the most suitable method for measuring the production of hydrogen.

....

.... **(2 marks)**

3 Which method would be most suitable for investigating the rate of this reaction?

$$CH_3COCH_3(aq) + I_2(aq) \rightarrow CH_3COCH_2I(aq) + HI(aq)$$

☐ **A** Measuring the volume of gas produced using a gas syringe.

☐ **B** Measuring the change in the mass using a balance.

☐ **C** Using a colorimeter to measure the light absorbance of the reaction mixture.

☐ **D** Titrating samples from the reaction mixture with acid. **(1 mark)**

4 Bromoethane reacts with sodium hydroxide solution to form ethanol and sodium bromide:

$$CH_3CH_2Br(aq) + NaOH(aq) \rightarrow CH_3CH_2OH(aq) + NaBr(aq)$$

The concentration of hydroxide ions was determined by titration with acid at regular intervals.

(a) State why ice-cold water should be added to each sample of the reaction mixture before titration.

.... **(1 mark)**

(b) Describe how you could calculate the initial rate of reaction from a concentration–time graph.

....

.... **(2 marks)**

Rate equation and initial rate

Guided   **1**   Describe the meaning of the term **order** in terms of a substance in a rate equation.

The order with respect to a substance in a rate equation is the ....

.... **(2 marks)**

Maths skills   **2**   The initial rate of the reaction between substances X and Y was measured in a series of experiments. This rate equation was determined: rate = $k[X][Y]^2$

(a) Complete the table below.

Experiment	Initial [X] /mol dm^{-3}	Initial [Y] /mol dm^{-3}	Initial rate /mol dm^{-3} s^{-1}
1	0.10	0.10	5.0×10^{-3}
2	0.40		2.0×10^{-2}
3	0.10	0.40	
4	0.02		9.0×10^{-3}

(3 marks)

(b) (i) Use the data from Experiment 1 to calculate a value for the rate constant, k.

.... **(2 marks)**

(ii) State the units for k.

.... **(1 mark)**

(c) State the change in the reaction conditions, if any, which would change the value of k.

.... **(1 mark)**

3   The following data were found for the reaction between substances A, B and C.

Experiment	Initial [A] /mol dm^{-3}	Initial [B] /mol dm^{-3}	Initial [C] /mol dm^{-3}	Initial rate /mol dm^{-3} s^{-1}
1	2.50×10^{-3}	3.75×10^{-3}	3.75×10^{-3}	0.90
2	1.25×10^{-3}	1.25×10^{-3}	1.25×10^{-3}	0.05
3	1.25×10^{-3}	2.50×10^{-3}	1.25×10^{-3}	0.20
4	2.50×10^{-3}	1.25×10^{-3}	1.25×10^{-3}	0.10

(a) Determine the orders of reaction with respect to A, B and C.

> Work out the orders with respect to A and B first, then use these with Experiments 1 and 4 to work out the order of reaction with respect to C.

....

.... **(3 marks)**

(b) Write the rate equation for the reaction.

.... **(1 mark)**

(c) Determine the overall order of reaction.

.... **(1 mark)**

Rate equation and half-life

1 Sketch labelled graphs of rate against the concentration of a reactant, where the order of reaction with respect to the reactant is zero order, first order and second order.

(3 marks)

2 In a decomposition reaction, the concentration of the reactant, A, was measured over time. The table shows the results.

(a) Plot a graph of [A] against t, and draw a line of best fit.

Time, t /s	[A] /mol dm^{-3}
0	0.078
60	0.058
120	0.044
180	0.033
240	0.025
300	0.018
360	0.014
420	0.010

(4 marks)

(b) Use your graph to determine the half-life, $t_{\frac{1}{2}}$, for the reaction starting from:

(i) [A] = 0.060 mol dm^{-3} ....

(ii) [A] = 0.030 mol dm^{-3} .... **(2 marks)**

(c) Explain, using your answers to part (b), the order of reaction with respect to reactant A.

....

.... **(2 marks)**

(d) Determine the instantaneous rate of reaction at 100 s.

> Draw a tangent to the curve.

....

.... **(3 marks)**

117

Rate-determining steps

1 Give the meaning of the term **rate-determining** step.

....

.... **(1 mark)**

2 2-bromo-2-methylpropane reacts with hydroxide ions to form 2-methylpropan-2-ol. The reaction proceeds in two steps:

Step 1: $(CH_3)_3CBr \rightleftharpoons (CH_3)_3C^+ + Br^-$

Step 2: $(CH_3)_3C^+ + OH^- \rightarrow (CH_3)_3COH$

(a) Write an equation for the overall reaction of 2-bromo-2-methylpropane with hydroxide ions.

.... **(1 mark)**

(b) The rate-determining step is Step 1.

(i) Suggest why Step 1, rather than Step 2, is the rate-determining step.

> Consider the bonds involved at each step, and the properties of the reacting species.

....

.... **(2 marks)**

(ii) The rate equation for the reaction is: rate = $k[(CH_3)_3Br]$
Explain why hydroxide ions do not appear in this equation.

....

.... **(2 marks)**

(c) Explain whether the reaction mechanism is an S_N1 mechanism or an S_N2 mechanism.

....

.... **(2 marks)**

3 Compound X reacts with compound Y as:

   $X + 3Y \rightarrow XY_3$

The rate equation for the reaction is: rate = $k[X][Y]^2$
The mechanism suggested for the reaction involves three steps:

Step 1: $X + Y \rightarrow XY$

Step 2: $XY + Y \rightarrow XY_2$

Step 3: $XY_2 + Y \rightarrow XY_3$

Explain which step (1, 2 or 3) is the rate-determining step.

> Compare the amount of X and Y in each step with the rate equation.

....

.... **(2 marks)**

Finding the activation energy

1 Hydrogen iodide decomposes to form hydrogen and iodine:

$$2HI(g) \rightarrow H_2(g) + I_2(g)$$

The table shows some experimental data about this reaction.

Temperature, T /K	Rate constant, k /mol dm^{-3} s^{-1}	$1/T$ /K^{-1}	$\ln k$
656	1.39×10^{-4}		
695	9.52×10^{-4}		
735	5.53×10^{-3}		
788	4.32×10^{-2}		
846	3.05×10^{-1}		

(a) Complete the table to show the values for $\ln k$ and $1/T$ (to three significant figures).

(2 marks)

(b) Plot a graph of $\ln k$ against $1/T$, and draw a line of best fit.

(4 marks)

(c) Calculate the gradient of the line.

....

....

.... **(2 marks)**

(d) The Arrhenius equation is: $k = A_e^{-E_a/RT}$
Use this equation to calculate the activation energy, E_a, for this reaction in kJ mol^{-1}.
Give your answer to three significant figures, and include the sign in your answer.
($R = 8.31$ J K^{-1} mol^{-1})

....

.... **(2 marks)**

Exam skills 11

1   Nitrogen dioxide decomposes when heated:

$$2NO_2(g) \rightarrow 2NO(g) + O_2(g)$$

The concentration of nitrogen dioxide was measured during a reaction.

(a)   Use the table of data to plot a graph of $[NO_2]$ against t, and draw a line of best fit.   **(4 marks)**

Time, t /s	$[NO_2]$ /mol dm^{-3}
0	1.00
50	0.79
100	0.65
150	0.55
200	0.48
250	0.43
300	0.38
350	0.34
400	0.31

(b)   Explain, using information from your graph, why the order of reaction with respect to NO_2 could be first order or second order, but not zero order.

....

....

.... **(3 marks)**

(c)   The instantaneous rate of reaction at different times can be determined from the graph plotted in part (a).

   (i)   Determine the instantaneous rate of reaction when $[NO_2] = 0.70$ mol dm^{-3}.

....

.... **(3 marks)**

   (ii)   Explain how you would use the method used in part (i) to determine if the reaction is second order with respect to NO_2.

....

....

.... **(3 marks)**

Identifying aldehydes and ketones

⟩Guided⟩ 1 (a) Complete the table to show the names of the compounds shown, and the observations expected in each situation.

Name of compound	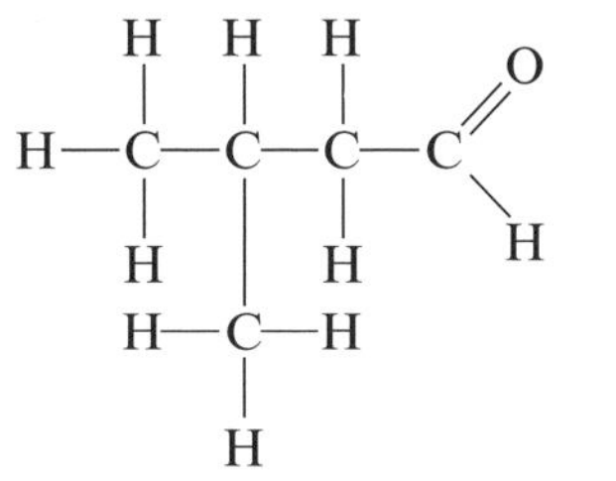 3-methyl	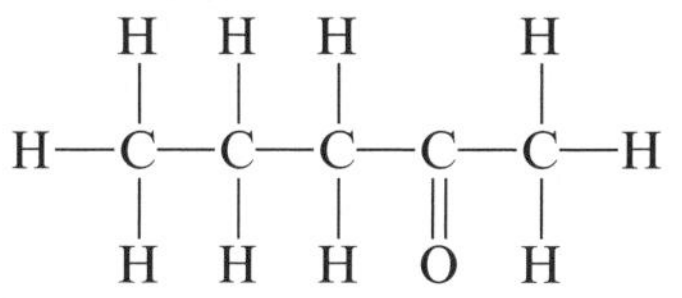 pentan ....
Warmed with potassium dichromate(VI) solution acidified with dilute H_2SO_4	....	....
Warmed with Fehling's solution	....	....
2,4-dinitrophenylhydrazine	....	....

(8 marks)

(b) Describe the differences, if any, between the observations seen when these two compounds are warmed with Benedict's solution and when they are warmed with Fehling's solution.

.... **(1 mark)**

2 Ethanal, CH_3CHO, gives a silver mirror when it is warmed with Tollens' reagent. What type of reaction is this?

☐ **A** Neutralisation

☐ **B** Redox

☐ **C** Addition

☐ **D** Substitution   **(1 mark)**

3 The table shows the boiling temperatures of three carbonyl compounds.

(a) Explain the trend in boiling temperature.

> Mention intermolecular forces in your answer.

Compound	Boiling temperature /°C
HCHO	−21
CH_3CHO	21
CH_3CH_2CHO	49

....

.... **(2 marks)**

(b) Explain why these aldehydes are soluble in water.

....

.... **(2 marks)**

121

Optical isomerism

1 Which of the following isomers of $C_3H_6Br_2$ shows optical activity?

> It may help to sketch their displayed formulae.

☐ **A** $CH_3CH_2CHBr_2$

☐ **B** $CH_3CHBrCH_2Br$

☐ **C** $CH_3CBr_2CH_3$

☐ **D** $CH_2BrCH_2CH_2Br$

(1 mark)

Maths skills

2 A compound X has the molecular formula, C_4H_9Cl.

(a) Give the displayed formula and name of the unbranched isomer of compound X which:

(i) does *not* show optical isomerism

Name: ....

(ii) does show optical isomerism.

Name: ....

(4 marks)

(b) Explain why the compound given in your answer to part (a)(ii) exists as optical isomers.

....

.... **(2 marks)**

(c) Explain the meaning of the term **racemic mixture**.

....

.... **(2 marks)**

(d) Explain the meaning of the term **plane-polarised monochromatic light**.

....

.... **(2 marks)**

3 Describe the essential features of using polarimetry to distinguish between two optical isomers of a given compound.

> Only include the important steps in the method, and do not include details of the apparatus or how it works

....

....

....

.... **(4 marks)**

Optical isomerism and reaction mechanisms

Guided

1 The diagram shows the displayed formula of 2-bromobutane. One optically active isomer of 2-bromobutane is reacted with hydroxide ions to form butan-2-ol:

$$CH_3CH_2CHBrCH_3 + OH^- \rightarrow CH_3CH_2CH(OH)CH_3 + Br^-$$

Why does the organic product form as a mixture of optical isomers?

☐ **A** The reaction proceeds via an S_N2 mechanism.

☐ **B** Butan-2-ol contains a chiral centre.

☐ **C** 2-bromobutane forms a five-bonded transition state.

☐ **D** 2-bromobutane forms a carbocation intermediate.

> Option **B** contains a correct statement but this does not explain the formation of a mixture of optical isomers.

(1 mark)

2 Ethanal reacts with HCN, in the presence of cyanide ions from KCN, to form 2-hydroxypropanenitrile.

> Treat the C=O bond as a single bond, and then count the number of electron pairs around the C atom.

(a) Predict the shape and bond angle around the carbon atom in the carbonyl group in ethanal.

.... **(2 marks)**

(b) Use your answer to part (a) to suggest how the cyanide ion, CN^-, can approach the $C^{\delta+}$ atom in the carbonyl group.

> Can it approach from one side, or more?

.... **(1 mark)**

(c) Use your answer to part (b) to explain why a racemic mixture of 2-hydroxypropanenitrile forms.

> A racemic mixture contains equal amounts of each optical isomer.

....

....

.... **(2 marks)**

3 The diagram shows the displayed formula of 2-iodobutane. An optically active isomer of 2-iodobutane reacts with hydroxide ions to form an optically active isomer of butan-2-ol. Why does the organic product form as a single optical isomer, rather than as a mixture of optical isomers?

☐ **A** The reaction proceeds via an S_N1 mechanism.

☐ **B** The reaction proceeds via an S_N2 mechanism.

☐ **C** One optical isomer of butan-2-ol forms via a more stable carbocation.

☐ **D** Only one species is involved in the rate-determining step.

(1 mark)

Had a go ☐ Nearly there ☐ Nailed it! ☐

Reactions of aldehydes and ketones

1 (a) Propanone reacts with iodine in the presence of sodium hydroxide, forming an insoluble yellow product. Which of the following is the formula of that product?

☐ **A** CH_3COCH_2I ☐ **B** CH_3COCHI_2

☐ **C** CH_3I ☐ **D** CHI_3 **(1 mark)**

(b) Complete the table to show which substance(s) will react with iodine in the presence of sodium hydroxide to produce a yellow precipitate. Placing a tick (✓) in each correct box.

Compound	Forms precipitate?
CH_3OH	
CH_3CH_2OH	
CH_3CHO	
CH_3CH_2CHO	
$CH_3COCH_2CH_2CH_3$	
$CH_3CH_2COCH_2CH_3$	

(3 marks)

2 Lithium tetrahydridoaluminate, $LiAlH_4$, can reduce carbonyl compounds to alcohols.

(a) Give a suitable solvent in which to carry out these reactions.

.... **(2 marks)**

(b) Write an equation to represent the reaction of $LiAlH_4$ with the following carbonyl compounds. Name the organic product formed in each reaction.

Represent the reducing agent as [H] in the equations.

(i) Butanone: ....

Name of organic product: ....

(ii) Butanal: ....

Name of organic product: .... **(4 marks)**

3 Ethanal reacts with HCN, in the presence of a catalyst of cyanide ions from KCN, to form 2-hydroxypropanenitrile, $CH_3CH(OH)CN$.

(a) Name the type of reaction mechanism involved.

.... **(1 mark)**

(b) Give the mechanism for the reaction.

Use curly arrows in your diagrams.

(3 marks)

Carboxylic acids

1   Butane molecules and ethanoic acid molecules have similar numbers of electrons, but butane is a gas at room temperature and ethanoic acid is a liquid. This is because ethanoic acid has:

☐ **A** greater London forces than butane

☐ **B** hydrogen bonding but butane does not

☐ **C** stronger ionic bonds than butane

☐ **D** stronger covalent bonds than butane.   **(1 mark)**

2   Carboxylic acids can be prepared by the oxidation of alcohols or aldehydes.

(a) A suitable oxidising agent is acidified potassium dichromate(VI).

(i) Name a suitable dilute acid to acidify the potassium dichromate(VI) solution.

....   **(1 mark)**

(ii) Name the aldehyde that could be oxidised to produce ethanoic acid.

....   **(1 mark)**

(iii) Write an equation to represent the production of ethanoic acid from ethanol.

> Represent the oxidising agent as [O] in the equation.

....   **(2 marks)**

(b) Carboxylic acids can also be produced by the hydrolysis of nitriles in the presence of dilute hydrochloric acid.

(i) Write an equation to represent the production of ethanoic acid from ethanenitrile, CH_3CN.

....   **(2 marks)**

(ii) Give the reaction condition necessary for this preparation.

....   **(1 mark)**

3   Write an equation for the reaction between propanoic acid and calcium carbonate, and name the salt produced.

Equation: ....

Name of salt: ....   **(2 marks)**

4   Carboxylic acids react with phosphorus(V) chloride, PCl_5, to produce acyl chlorides.

(a) Write an equation to represent the reaction between ethanoic acid and PCl_5.

....   **(2 marks)**

(b) Suggest a method for removing the phosphorus-containing product from the organic product.

....   **(1 mark)**

Making esters

1 Esters may be synthesised by reacting an alcohol with a carboxylic acid or with an acyl chloride. Why does the reaction between an alcohol and an acyl chloride give a better yield of ester?

☐ **A** The reaction between an alcohol and a carboxylic acid needs an acid catalyst.

☐ **B** The reaction between an alcohol and a carboxylic acid goes to completion.

☐ **C** The reaction between an alcohol and an acyl chloride goes to completion.

☐ **D** The reaction between an alcohol and an acyl chloride produces misty fumes.   **(1 mark)**

2 The diagram shows the displayed formula for ethyl methanoate.

$$\begin{array}{c}\quad\ \ H\ \ \ H\qquad\quad O\\ \quad\ \ |\quad\ \ |\qquad\quad ||\\ H-C-C-O-C-H\\ \quad\ \ |\quad\ \ |\\ \quad\ \ H\ \ \ H\end{array}$$

  (a) This ester can be synthesised using an alcohol and a carboxylic acid, or using the same alcohol and an acyl chloride.

     (i) Name the alcohol and carboxylic acid needed.

     .... **(2 marks)**

     (ii) Write an equation to represent the reaction between the two substances in part (i).

     .... **(1 mark)**

     (iii) Name the acyl chloride needed to synthesise the ester using the alcohol in part (a)(i).

     .... **(1 mark)**

     (iv) Name the inorganic compound formed when an acyl chloride reacts with an alcohol.

     .... **(1 mark)**

  (b) Name, and draw the displayed formula for, the following isomers of methyl ethanoate.

	(i) **Another ester**	(ii) **A carboxylic acid**
Name of compound	....	....
Displayed formula		

                                                                 (4 marks)

3 (a) Esters can be hydrolysed under reflux conditions.

     (i) Define the term **hydrolysis**.

     .... **(1 mark)**

     (ii) Write an equation to show the acid hydrolysis of ethyl propanoate, $CH_3CH_2COOCH_2CH_3$.

     .... **(2 marks)**

  (b) Write an equation to show the hydrolysis of ethyl propanoate using sodium hydroxide solution.

  .... **(1 mark)**

Making polyesters

1 Give a similarity and a difference between addition polymerisation and condensation polymerisation.

Similarity: .... **(1 mark)**

Difference: .... **(1 mark)**

2 The repeat unit for a polyester is shown on the right.
Which two monomers could form this polymer?

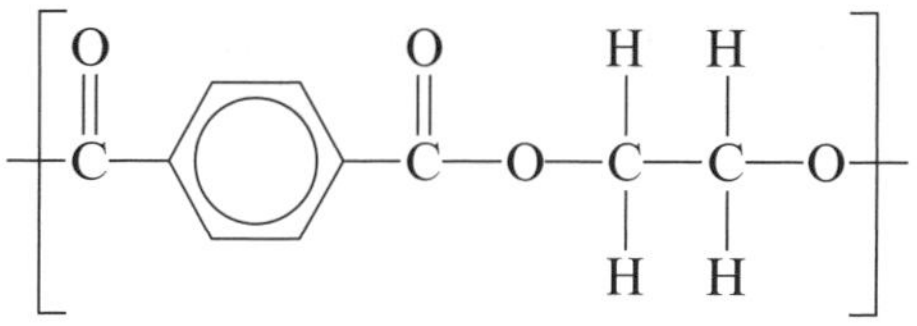

	Monomer 1	Monomer 2
☐ **A**	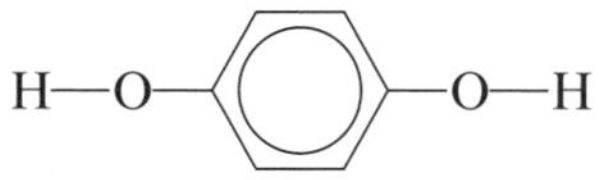	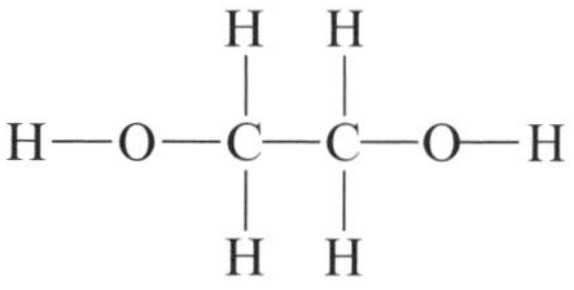
☐ **B**	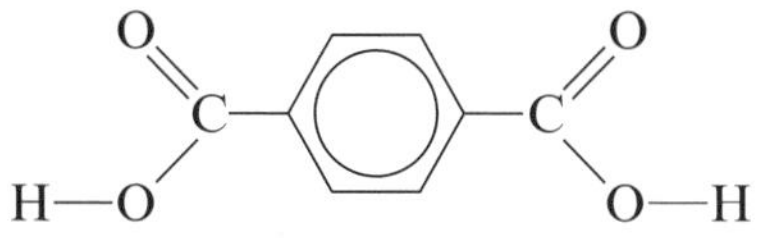	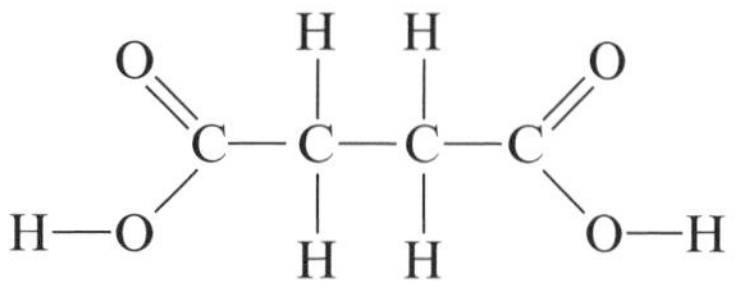
☐ **C**	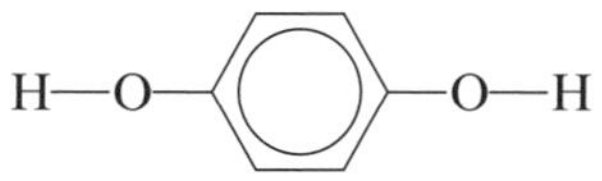	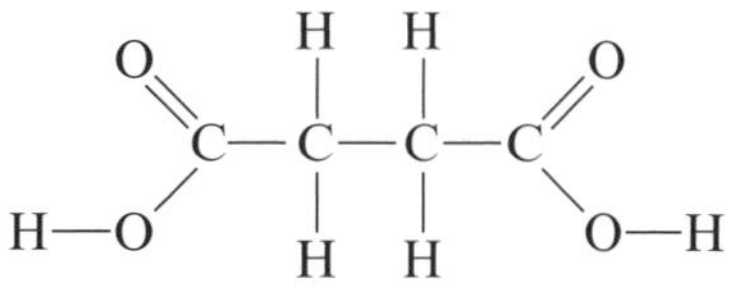
☐ **D**	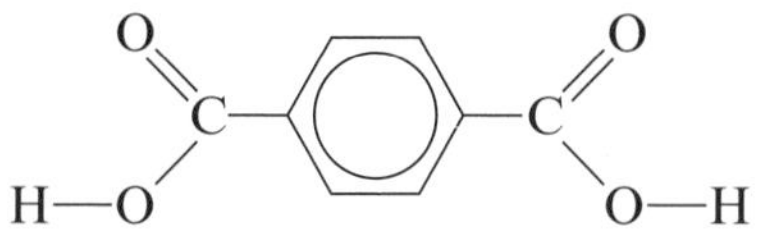	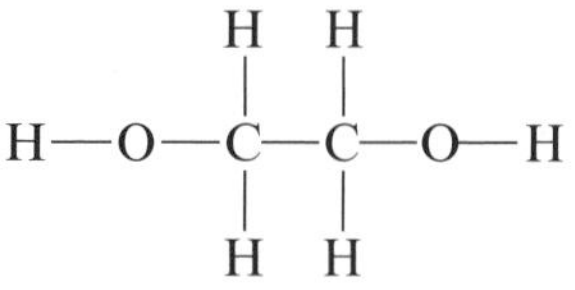

(1 mark)

3 The polymer called polybutylene succinate, or PBS, is a biodegradable polyester.
It is made from two monomers, shown below with their common names.

Displayed formula	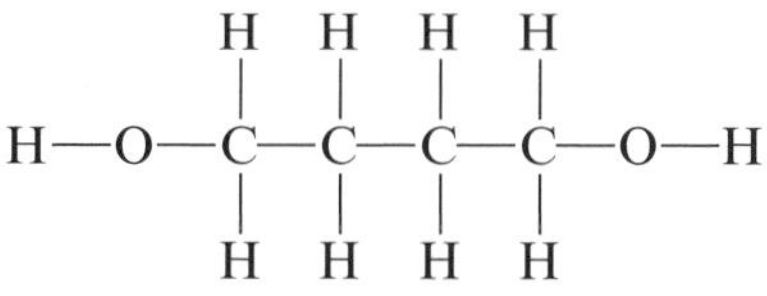	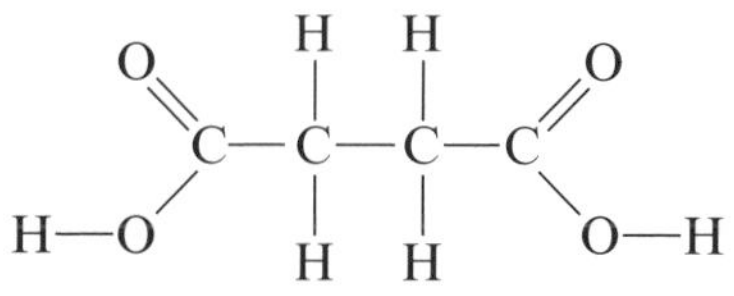
Common name	butylene glycol	succinic acid
Systematic name	....	....

(a) Complete the table to show the systematic (IUPAC) names for these two compounds. **(2 marks)**

(b) Draw the displayed formula for the repeat unit of PBS in the space below. **(2 marks)**

Benzene

1   The Kekulé structure for benzene is shown on the right.

   (a)   Write the molecular formula of benzene.

....   **(1 mark)**

   (b)   Benzene reacts with oxygen in the air.

      (i)   State what you would observe when benzene burns in air.

> In your answer, describe what the flame would look like.

....   **(1 mark)**

      (ii)   Write an equation to represent the incomplete combustion of benzene to produce water, and equal amounts of carbon, and carbon monoxide and carbon dioxide.

....   **(2 marks)**

   (c)   Benzene reacts with bromine under reflux in the presence of a catalyst. Name a suitable catalyst for the reaction.

....   **(1 mark)**

2   Benzene molecules contain π bonds.

   (a)   Describe how a π bond forms.

> Include the orbitals involved and how they overlap.

....

....

....   **(3 marks)**

   (b)   State the number of π bonds in a benzene molecule.

....   **(1 mark)**

3   The standard enthalpy change of hydrogenation of C=C bond is $-120\,\text{kJ mol}^{-1}$.

   (a)   Calculate the standard enthalpy change of hydrogenation for the Kekulé structure of benzene.

....   **(1 mark)**

   (b)   The measured value for the standard enthalpy change of hydrogenation for benzene is $-208\,\text{kJ mol}^{-1}$.

      (i)   Calculate the difference between the stability of benzene and the Kekulé structure for benzene.

> Use the information given and your answer to part (a).

....   **(1 mark)**

      (ii)   State what your answer to part (i) tells you about the bonding in benzene.

....   **(1 mark)**

Halogenation of benzene

1 The diagrams show the Kekulé structure of benzene and the structure of cyclohexa-1,3-diene.

benzene   cyclohexa-1,3-diene

Bromine reacts with cyclohexa-1,3-diene in the cold. It reacts with benzene under reflux conditions in the presence of a catalyst.

(a) Which of the following correctly describes the reaction mechanisms involved.

		Reaction of bromine with benzene	Reaction of bromine with cyclohexa-1,3-diene
☐	A	electrophilic addition	electrophilic addition
☐	B	electrophilic substitution	electrophilic addition
☐	C	electrophilic substitution	electrophilic substitution
☐	D	electrophilic addition	electrophilic substitution

(1 mark)

(b) Explain why benzene is resistant to bromination but alkenes are not.

....

.... **(2 marks)**

2 Benzene reacts with bromine in the presence of an $AlBr_3$ catalyst:

$$C_6H_6 + Br_2 \rightarrow C_6H_5Br + HBr$$

(a) (i) Give the formula of the electrophile generated by the reaction of $AlBr_3$ with Br_2.

.... **(1 mark)**

(ii) Write an equation to represent the formation of the species identified in part (i).

.... **(1 mark)**

(iii) Write an equation to show how the catalyst is regenerated from an ion in part (ii).

.... **(1 mark)**

(b) Give the reaction mechanism for the reaction between benzene and the electrophile.

(3 marks)

3 Explain why phenol, C_6H_5OH, decolourises bromine water.

> What effect does the hydroxyl group have?

....

.... **(3 marks)**

Nitration of benzene

1 Nitrobenzene is prepared using a mixture of two acids at 50 °C.

  (a) Name the two acids needed.

  .... **(2 marks)**

  (b) The nitration of benzene is an example of an electrophilic substitution reaction.

   (i) Give the formula of the electrophile involved.

   .... **(1 mark)**

   (ii) Write an equation to represent the formation of this electrophile from the two acids named in part (a).

   .... **(1 mark)**

   (iii) Give the reaction mechanism for the reaction between benzene and the electrophile.

   (3 marks)

2 1,3,5-trinitrobenzene, shown, is used in explosives for commercial mines.

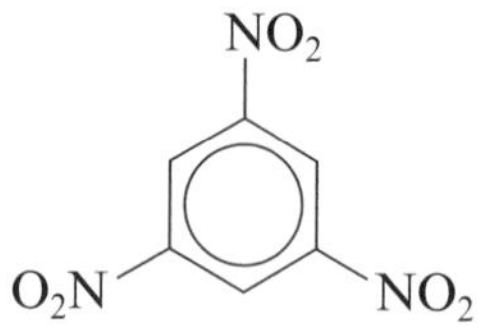

$$NO_2$$

$$O_2N \qquad NO_2$$

  It can be synthesised by the nitration of 1,3-dinitrobenzene.

  (a) Give the structure of 1,3-dinitrobenzene.

  (1 mark)

  (b) Give the structures of the two position isomers of 1,3-dinitrobenzene, and give their names.

Structure		
Name		

  (4 marks)

  (c) Suggest why nitration becomes increasingly difficult as each successive nitro group is added to benzene.

  ┌─────────────────────────────┐
  │ What effect will the nitro group │
  │ have on the delocalised π bond? │
  └─────────────────────────────┘

  ....
  .... **(2 marks)**

Friedel–Crafts reactions

Guided 1 The diagram represents a molecule of phenylpropanone.

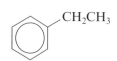

Benzene reacts with propanoyl chloride, in the presence of an aluminium chloride catalyst, to form phenylpropanone and hydrogen chloride.

(a) Write two equations to show aluminium chloride acting as a catalyst in the reaction.

> When you show the electrophile, place the + against the carbon atom where the C–Cl bond has broken.

$CH_3CH_2COCl + AlCl_3 \rightarrow$ ....

$[AlCl_4]^- +$ .... **(2 marks)**

(b) Give the reaction mechanism for the reaction between benzene and the electrophile.

(3 marks)

(c) Aluminium chloride accepts a pair of electrons as part of its function as a catalyst in the reaction between benzene and propanoyl chloride.

(i) State the feature of aluminium chloride that allows it to do this.

> Think about the arrangement of outer electrons in the aluminium atom.

.... **(1 mark)**

(ii) Explain whether ammonium chloride could also catalyse the reaction.

....

.... **(2 marks)**

2 The diagram represents a molecule of ethylbenzene.

CH_2CH_3

Benzene reacts with chloroethane, in the presence of an aluminium chloride catalyst, to form ethylbenzene.

(a) Describe the reaction conditions necessary for this reaction.

.... **(2 marks)**

(b) Write the reaction mechanism for the reaction between benzene and the electrophile formed.

(3 marks)

Making amines

1 Butan-2-amine, $CH_3CH_2CH(NH_2)CH_3$, can be synthesised by reacting concentrated ammonia with:

☐ **A** butan-2-ol

☐ **B** 2-bromobutane

☐ **C** butane

☐ **D** but-2-ene

(1 mark)

2 State whether each compound in the table below is a primary amine, secondary amine or tertiary amine.

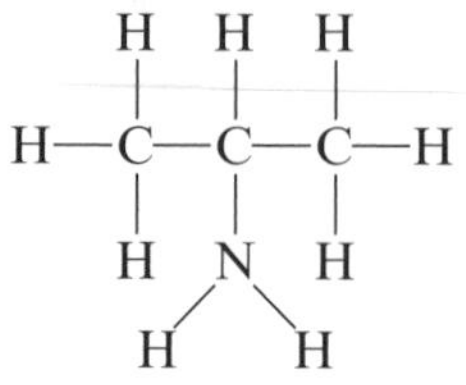	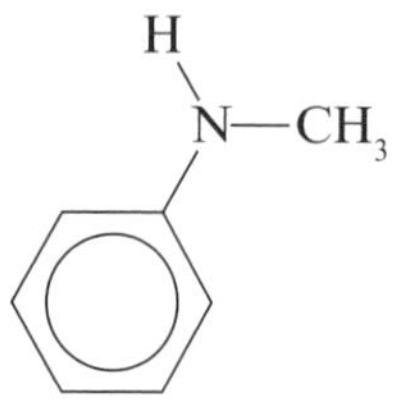	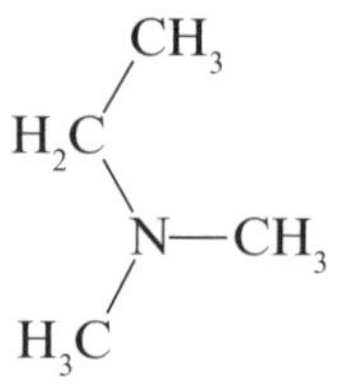	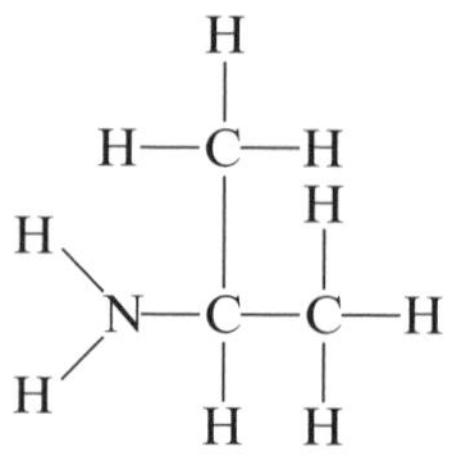
....	....	....	....

(4 marks)

3 Ethylamine, $CH_3CH_2NH_2$, can be produced by reducing ethanenitrile, CH_3CN, using $LiAlH_4$.

(a) Give the name for $LiAlH_4$.

.... **(1 mark)**

(b) Write an equation for the reaction. | Show the reducing agent as [H]. |

.... **(1 mark)**

4 Nitrobenzene can be reduced to phenylamine using tin and concentrated hydrochloric acid.

(a) State the conditions under which the reaction mixture is heated.

.... **(1 mark)**

(b) Write an equation for the reaction. | Show the reducing agent as [H]. |

.... **(2 marks)**

5 Ethylamine can be produced by heating ethanol with ammonia in a sealed container.

(a) Write an equation for the overall reaction that occurs.

.... **(1 mark)**

(b) Explain why excess ammonia is needed in this organic synthesis.

| Ammonia and amines have a lone pair of electrons on their nitrogen atoms. |

....

....

.... **(3 marks)**

Amines as bases

1 Ammonia, NH_3, phenylamine, $C_6H_5NH_2$, and ethylamine, $CH_3CH_2NH_2$, all dissolve in water. Which of the following shows the pH of $0.1 \, mol \, dm^{-3}$ solutions in order of decreasing pH?

☐ **A** ammonia, ethylamine, phenylamine

☐ **B** phenylamine, ammonia, ethylamine

☐ **C** ethylamine, phenylamine, ammonia

☐ **D** ethylamine, ammonia, phenylamine   **(1 mark)**

2 Ammonia can act as a Brønsted–Lowry base.

(a) State what is meant by the term **Brønsted–Lowry base**.

.... **(1 mark)**

(b) Explain, with the help of an equation, why ammonia acts as a Brønsted–Lowry base in solution.

....

.... **(2 marks)**

3 Amines can act as Brønsted–Lowry bases.

(a) State the feature of an amine molecule that allows it to act as a Brønsted–Lowry base.

.... **(1 mark)**

(b) Explain the difference in base strength between primary aliphatic amines such as methylamine, CH_3NH_2, and ammonia, NH_3.

> Say what the difference is, then explain it.

....

.... **(3 marks)**

(c) Explain the difference in base strength between aromatic amines such as phenylamine, $C_6H_5NH_2$, and ammonia, NH_3.

> Say what the difference is, then explain it.

....

.... **(3 marks)**

4 Propylamine, $C_3H_7NH_2$, dissolves in water to form an alkaline solution.

(a) State why, in terms of intermolecular forces, propylamine is soluble in water.

.... **(1 mark)**

(b) Write an equation for the reaction of propylamine with water.

.... **(1 mark)**

(c) Write an equation for the reaction of propylamine with hydrochloric acid. Name the product.

.... **(2 marks)**

Other reactions of amines

1 Ammonia solution is added to copper(II) sulfate solution. Which of the following observations would be made?

☐ **A** A blue precipitate forms which dissolves to form a deep blue solution when excess ammonia solution is added.

☐ **B** A blue precipitate forms which remains when excess ammonia solution is added.

☐ **C** A white precipitate forms which dissolves to form a colourless solution when excess ammonia solution is added.

☐ **D** A white precipitate forms which remains when excess ammonia solution is added.

(1 mark)

2 Propylamine, $C_4H_9NH_2$, reacts with hexaaquacopper(II) ions, $[Cu(H_2O)_6]^{2+}$, in aqueous solution.

(a) (i) Write an equation, including state symbols, to show the reaction between propylamine and hexaaquacopper(II) ions to form a precipitate.

.... **(3 marks)**

(ii) Name the type of reaction shown in part (i).

.... **(1 mark)**

(b) (i) Write an equation, including state symbols, to show the reaction between excess propylamine and the precipitate formed in the reaction in part (a).

.... **(3 marks)**

(ii) Name the type of reaction shown in part (i).

.... **(1 mark)**

3 The diagram represents *N*-methylethylamine.

(a) *N*-methylethylamine is produced in the reaction between methylamine, CH_3NH_2, and a chloroalkane.

(i) Name the chloroalkane required, and give its formula.

.... **(2 marks)**

(ii) Write an equation to show the reaction between methylamine and the chloroalkane.

.... **(1 mark)**

(b) State the type of amine to which *N*-methylethylamine belongs.

> Is it primary, secondary or tertiary?

.... **(1 mark)**

(c) Explain why *N*-ethyl-*N*-methylethylamine also forms in the reaction mixture in part (a).

....

.... **(3 marks)**

Making amides

1   Amides can be made in reactions with methanoyl chloride, HCOCl. Which of the following will not form an amide in a reaction with methanoyl chloride?

    ☐ **A** $(CH_3)_3N$

    ☐ **B** $(CH_3)_2NH$

    ☐ **C** CH_3NH_2

    ☐ **D** NH_3 **(1 mark)**

2   This molecule contains several functional groups.

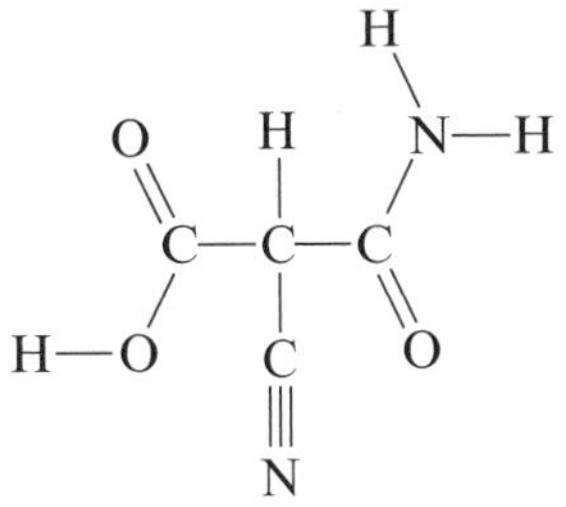

    (a)   Identify the nitrile functional group by drawing a ring around it.
Label this A.   **(1 mark)**

    (b)   Identify the amide functional group by drawing a ring around it.
Label this B.   **(1 mark)**

3   This diagram represents an acyl chloride.

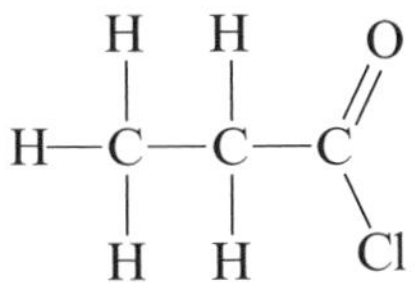

    (a)   Give the structural formula and name of this acyl chloride.

    .... **(2 marks)**

    (b)   Write an equation for the reactions between this acyl chloride and the following compounds.

       (i)   concentrated ammonia

    .... **(1 mark)**

       (ii)   ethylamine

    .... **(1 mark)**

    (c)   Name the organic product of the reaction shown in part (b)(i).

> The product is named after its number of carbon atoms, ending in 'amide'.

             (1 mark)

    ....

4   This diagram represents a secondary amide.

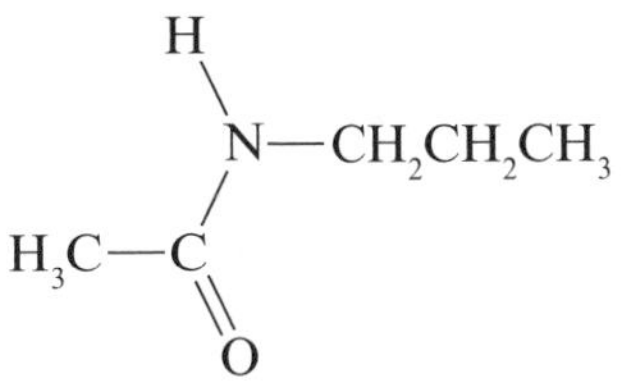

    (a)   Explain why this compound is a secondary amide.

    ....

    .... **(2 marks)**

    (b)   Explain why this compound is called *N*-propylethanamide.

> Look at the alkyl group attached to the nitrogen atom, and the amide group.

    ....

    .... **(3 marks)**

Making polyamides

1 The repeat unit for a polyamide is shown.

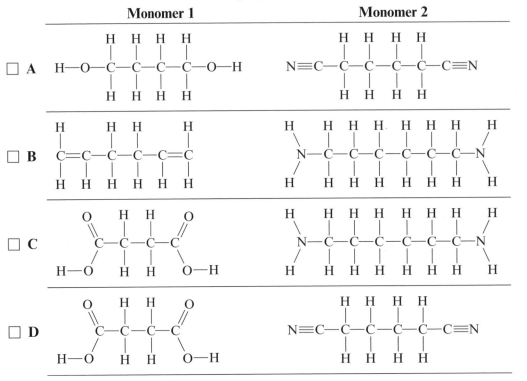

Which two monomers could form this polymer?

	Monomer 1	Monomer 2
☐ A	H—O—C—C—C—C—O—H (with H's)	N≡C—C—C—C—C—C≡N (with H's)
☐ B	C=C—C—C—C=C (with H's)	N—C—C—C—C—C—C—N (with H's)
☐ C	O=C—C—C—C=O with H—O and O—H	N—C—C—C—C—C—C—N (with H's)
☐ D	O=C—C—C—C=O with H—O and O—H	N≡C—C—C—C—C—C≡N (with H's)

(1 mark)

2 Polyphthalamides are tough condensation polymers made from two monomers. Two examples of these monomers are shown below with their common names.

Displayed formula	H_2N—C—C—C—C—NH_2 (with H's)	(benzene ring with C=O and Cl groups at 1,4 positions)
Common name	putrescine	terephthaloyl chloride
Systematic name	....	benzene-1,4-dicarbonyl dichloride

(a) Complete the table to show the systematic (IUPAC) name for putrescine. **(1 mark)**

(b) Suggest why one of the monomers is a dioyl dichloride rather than a dicarboxylic acid.

.... **(1 mark)**

(c) Draw the displayed formula for the repeat unit of the polyphthalamide formed by these monomers.

(2 marks)

Amino acids

1 The displayed formula for valine, an amino acid, is shown.

> The carbon atoms of the carboxyl group has the locant 1.

(a) Give the systematic name for valine.

.... **(1 mark)**

(b) Explain why valine is an amino acid.

.... **(2 marks)**

(c) Explain whether or not valine has optical isomers.

> Determine whether the valine molecule has an asymmetric carbon atom.

....

.... **(3 marks)**

Guided 2 Butylamine, $CH_3(CH_2)_3NH_2$, propanoic acid, CH_3CH_2COOH, and aminoethanoic acid (glycine), NH_2CH_2COOH, have similar molar masses. Butylamine and propanoic acid are in the liquid state at room temperature, but glycine is in the solid state. This is mainly because:

☐ ~~A~~ glycine has optical isomers but butylamine and propanoic acid do not

☐ **B** glycine has stronger London forces between its molecules

☐ **C** glycine has stronger hydrogen bonds between its molecules

☐ **D** glycine has strong ionic bonds between its molecules in the solid state. **(1 mark)**

> Option **A** cannot be correct because glycine does not have optical isomers.

3 The structure of 2-aminopropanoic acid (alanine) is shown.

(a) Complete the table below to show the structure of alanine in acidic solution and in alkaline solution.

Structure		

  (i) **acidic solution**   (ii) **alkaline solution** **(2 marks)**

(b) With reference to amino acid molecules, explain what is meant by the term **zwitterion**.

....

.... **(2 marks)**

(c) Write an equation to represent the reaction of alanine, $NH_2CH(CH_3)COOH$, with potassium hydroxide solution.

.... **(1 mark)**

Proteins

> **Guided**

1 Proteins can be hydrolysed to produce a mixture of their constituent amino acids. Which of the following is the best method for separating a mixture of amino acids in solution?

☐ **A** Filtration

☐ **B** Chromatography

> Option **D** cannot be correct because infrared spectroscopy is an analytical technique not a separation method.

☐ **C** Fractional distillation

☐ **D̸** Infrared spectroscopy

(1 mark)

2 The displayed formula for the amino acid glycine, aminoethanoic acid, is shown.
Two glycine molecules may react together to form a dipeptide.

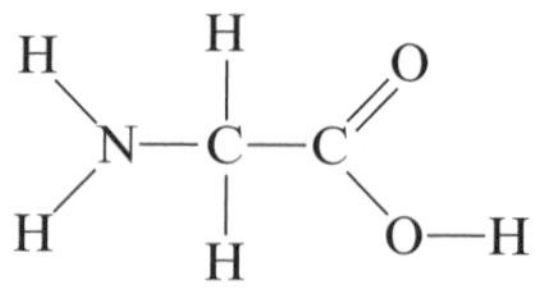

(a) Give the displayed formula of the dipeptide.

(1 mark)

(b) Identify the peptide group by drawing a box around it on your diagram in part (a).   **(1 mark)**

(c) Many glycine molecules may react together to form polyglycine, i.e. poly(aminoethanoic acid).

(i) Give the repeat unit for this polypeptide.

(2 marks)

(ii) State the type of polymerisation involved in forming this polypeptide.

....   **(1 mark)**

3 Proteins may be hydrolysed under acidic conditions.

(a) Describe the conditions necessary for this acidic hydrolysis.

> Include the acid needed.

....

....   **(3 marks)**

> **Guided**

(b) The displayed formula for the amino acid serine,
2-amino-3-hydroxypropanoic acid, is shown.
Serine has this structure: $H_2NCH(CH_2OH)COOH$.
Serine forms a dipeptide with this structure:
$H_2NCH(CH_2OH)CONHCH(CH_2OH)COOH$.
Write an equation to represent the acid hydrolysis
of this dipeptide.

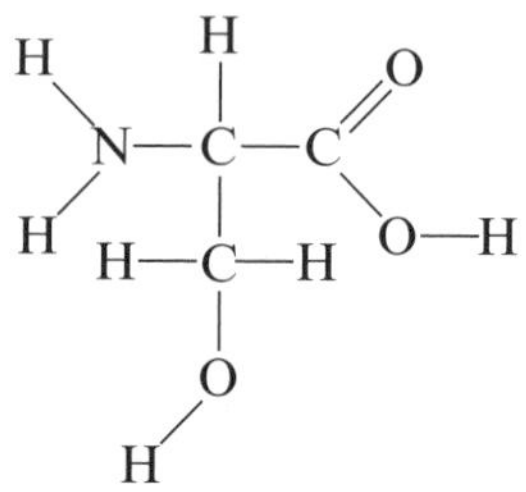

> Show the acid as H+. You will need three reactant species and one product.

$H_2NCH(CH_2OH)CONHCH(CH_2OH)COOH$ + ...

....   **(2 marks)**

Exam skills 12

1 Butylamine, $CH_3(CH_2)_3NH_2$, can be prepared from 1-chloropropane, $CH_3CH_2CH_2Cl$, in two steps.

(a) Write an equation for each step.

Step 1: ....

Step 2: .... **(2 marks)**

(b) Name the organic product of the first step.

.... **(1 mark)**

2 Phenylamine, $C_6H_5NH_2$, is slightly soluble in water, forming a weakly alkaline solution.

(a) Explain the feature of phenylamine that allows it to act as a base.

....

.... **(2 marks)**

(b) Explain which is the stronger base, phenylamine or ammonia.

....

....

.... **(3 marks)**

(c) Write an equation for the reaction that occurs when phenylamine is dissolved in water, and explain the role of water in this reaction.

....

.... **(3 marks)**

3 The diagrams show the structures of two amino acids, glycine and alanine.

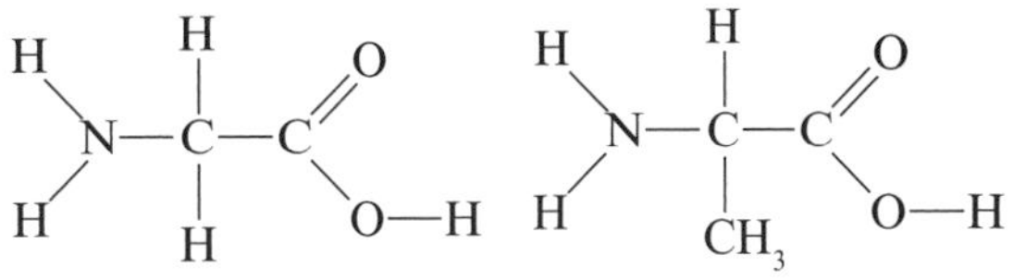

glycine   alanine

(a) Give the systematic (IUPAC) name for alanine.

.... **(1 mark)**

(b) Draw two different dipeptides that could be formed from a molecule of glycine and a molecule of alanine.

Dipeptide 1	Dipeptide 2

(2 marks)

Grignard reagents

1 Grignard reagents are organometallic compounds with a carbon–metal bond.

(a) Name the metal found in Grignard reagents.

.... **(1 mark)**

(b) Name the type of organic compound that reacts with the metal named in part (a) to form a Grignard reagent.

.... **(1 mark)**

(c) Give the reaction condition and solvent needed for the reaction described in part (b).

.... **(2 marks)**

(d) State why the solvent must be dry.

.... **(1 mark)**

2 If the desired product is a primary alcohol, a Grignard reagent and methanal are required.
Complete the table below to show the other reactant needed to produce each type of compound.

Desired product	Grignard reagent with
primary alcohol	methanal
secondary alcohol	....
tertiary alcohol	....
carboxylic acid	....

(3 marks)

3 Butan-1-ol, $CH_3CH_2CH_2CH_2OH$, can be made from methanal, HCHO, and the Grignard reagent formed from 1-bromopropane, $CH_3CH_2CH_2Br$, and magnesium.

(a) Write an equation to show the formation of the Grignard reagent.

.... **(1 mark)**

(b) Write an equation to show the reaction of the Grignard reagent with methanal.

.... **(1 mark)**

(c) Write an equation to show the hydrolysis, in acidic conditions, of the product formed in part (b) to produce butan1-ol.

> You do not need to show the acid in the equation.

.... **(2 marks)**

(d) Describe the differences between the organic compound used to form the Grignard reagent and the final organic product shown in part (c).

.... **(2 marks)**

Methods in organic chemistry 1

1 Aspirin, 2-ethanoyloxybenzoic acid, can be made from salicylic acid, 2-hydroxybenzoic acid:

salicylic acid + ethanoic anhydride → aspirin + ethanoic acid

A few drops of concentrated sulfuric acid are added to act as a catalyst.
The reaction is allowed to continue in a hot water bath for about 20 minutes at 60 °C, then ice-cold water is added to produce a precipitate of impure aspirin.

(a) State a hazard of the catalyst. Other than eye protection, explain a suitable control measure.

....

....

.... **(3 marks)**

(b) The impure aspirin crystals are separated from the reaction mixture by filtration under reduced pressure, and washed with ice-cold water.

Suggest an advantage of using filtration under reduced pressure, rather than sedimentation or filtration under gravity, to separate the crystals from the reaction mixture.

impure aspirin

filter paper

Buchner funnel

to pump

....

.... **(1 mark)**

(c) The impure aspirin crystals are dissolved in a minimum volume of hot ethyl ethanoate, and the mixture is then filtered while it is still hot. The filtrate is cooled in an ice bath to produce aspirin crystals, which are then removed by filtration and dried.

This describes recrystallisation.

(i) State why a 'minimum volume of hot ethyl ethanoate' is used.

....

(ii) State how insoluble impurities are removed in this process.

....

....

(iii) State how soluble impurities are removed in this process.

....

.... **(3 marks)**

(d) Describe how you could use melting point data to determine the purity of the aspirin.

....

.... **(2 marks)**

Had a go ☐ Nearly there ☐ Nailed it! ☐

Methods in organic chemistry 2

1 Nitrobenzene, $C_6H_5NO_2$, can be reduced to form phenylamine, $C_6H_5NH_2$, using tin in concentrated hydrochloric acid. Which method can be used to separate phenylamine from the reaction mixture?

☐ **A** Recrystallisation

☐ **B** Filtration

☐ **C** Steam distillation

☐ **D** Chromatography **(1 mark)**

Practical skills

2 Ethanal can be produced by the oxidation of ethanol. Ethanol is mixed with acidified potassium dichromate(VI), and the mixture is heated under simple distillation conditions.

(a) Draw a labelled diagram to show the apparatus needed to carry out this synthesis.

(3 marks)

(b) If the reaction mixture is heated under reflux conditions instead, the product is ethanoic acid not ethanal. Draw a labelled diagram to show the apparatus needed to carry out this synthesis.

(3 marks)

3 Fragrant oils, such as lavender oil, are extracted from crushed plant material using steam distillation.

(a) Describe how steam distillation is carried out.

....

.... **(3 marks)**

(b) Give an advantage of steam distillation over normal distillation.

.... **(1 mark)**

Reaction pathways

1 Ethanoic acid can be produced from ethene in two steps:

For each step, give the necessary reagents and conditions required.

Step 1: ....

.... **(3 marks)**

Step 2: ....

.... **(3 marks)**

2 Phenylethene (styrene) is the monomer used to make the addition polymer called polystyrene. Phenylethene can be produced from benzene in three steps:

For each step, give the necessary reagents and conditions required.

> Step 1 is a Friedel–Crafts acylation reaction.

Step 1: ....

.... **(3 marks)**

> Steps 2 and 3 do not involve the benzene ring.

Step 2: ....

.... **(2 marks)**

Step 3: ....

.... **(2 marks)**

3 1,2-dichloroethane, CH_2ClCH_2Cl, can be made by the reaction of ethene with chlorine, or by the reaction of ethane with chlorine. Suggest an advantage and a disadvantage of each method.

> Consider the nature of the reactants and products, and the atom economies.

....

....

.... **(4 marks)**

Chromatography

1 The diagram shows the results of thin layer chromatography, TLC.

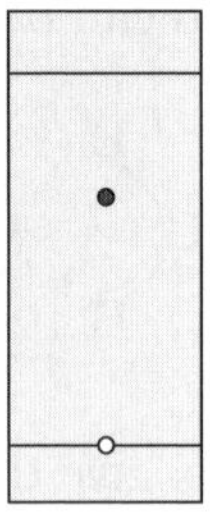

   (a) Determine the distance travelled from the base line by:

   (i) the solvent: ....

   ....

   (ii) the sample spot: .... **(1 mark)**

   (b) Use your answers to part (a) to calculate the R_f value for the sample, to two significant figures.

   .... **(1 mark)**

2 Sudan I is a red dye that is not permitted for use in food. Food scientists suspected that packets of chilli powder have been contaminated with Sudan I. They tested samples of three different chilli powders using gas chromatography, GC. For comparison, they also tested a known stock chilli powder and a sample of Sudan I. Their results are shown here. Use these chromatograms to answer the questions.

   From database for comparison

   stock chilli powder

   Sudan I

   Chilli powders being tested

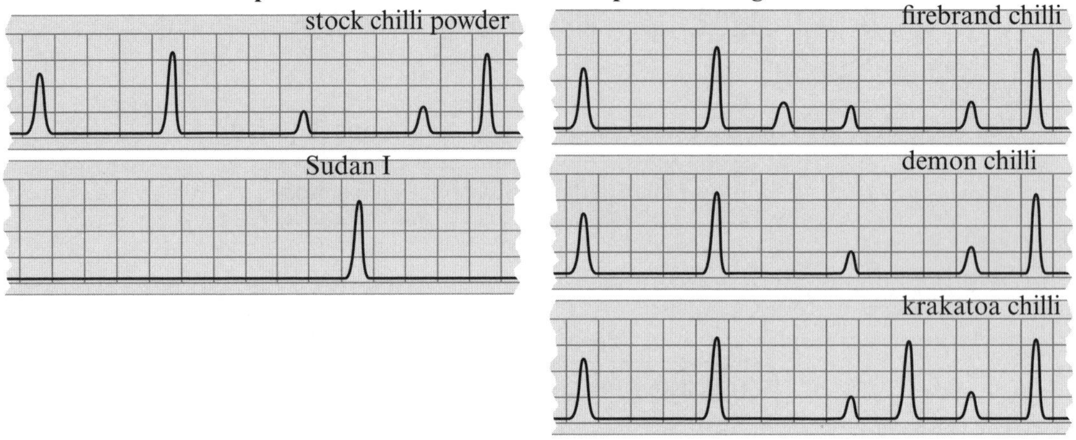

   firebrand chilli

   demon chilli

   krakatoa chilli

   (a) Explain whether or not the stock chilli powder contained Sudan I.

   ....

   .... **(2 marks)**

   (b) Explain which chilli powder was contaminated with Sudan I.

   ....

   .... **(3 marks)**

   (c) The food scientists assumed that their sample of Sudan I was pure.

   (i) Identify their evidence for this assumption.

   .... **(1 mark)**

   (ii) Suggest a further analytical method that the food scientists could use in conjunction with GC to further analyse the sample of Sudan I.

   .... **(1 mark)**

Functional group analysis

> **Guided**

1 'Iodoform', CHI_3, is an insoluble yellow solid that forms in reactions between certain organic compounds and an alkaline solution of iodine. Which of these compounds will produce iodoform?

   ☐ **A** $(CH_3)_3OH$

   ☐ **B** $CH_3CH_2CH_2OH$

   ☐ **C̸** CH_3CH_2CHO

   ☐ **D** CH_3COCH_3

> Option **C** cannot be correct because ethanal is the only aldehyde to give a positive result in the iodoform test.

(1 mark)

> **Practical skills**

2 Describe how you could use one simple test tube reaction to distinguish between the compounds in the following pairs. In each answer, state a suitable reagent, and describe the observations you expect for each compound.

(a) hex-1-ene and hexane

....

....

.... **(3 marks)**

> **Guided**

(b) 2-methylpropan-2-ol and 2-methylpropan-1-ol

Add $K_2Cr_2O_7$ acidified with ....

....

.... **(3 marks)**

(c) propanoic acid and methyl ethanoate

....

.... **(3 marks)**

(d) propanal and propanone.

....

.... **(3 marks)**

3 1-chloropropane, 1-bromopropane and 1-iodopropane are liquids at room temperature.
Describe how you would distinguish between them using simple test tube reactions.

> Describe how you could release halide ions from these compounds, and then identify the halide ions released.

....

....

....

.... **(7 marks)**

Combustion analysis

1 (a) Write an equation for the complete combustion of methane, CH_4.

.... **(1 mark)**

(b) $50\,cm^3$ of methane undergoes complete combustion in oxygen. Assuming all volumes of gases are measured at the same temperature and pressure, calculate:

(i) the minimum volume of oxygen required

.... **(1 mark)**

(ii) the volume of carbon dioxide formed.

.... **(1 mark)**

2 A $25\,cm^3$ sample of a gaseous hydrocarbon required exactly $200\,cm^3$ of oxygen for complete combustion. If all volume measurements were made at the same temperature and pressure, which of the following is the correct formula for the hydrocarbon?

☐ **A** C_4H_8

☐ **B** C_4H_{10}

☐ **C** C_5H_{10}

☐ **D** C_5H_{12} **(1 mark)**

⟩**Guided**⟩ **3** Compound Z consists of carbon, hydrogen and oxygen only. A sample was completely burnt in oxygen. The table shows the results of the experiment.

Substance	Mass /g
compound Z	2.25
carbon dioxide	4.39
water	2.25

(a) Calculate the mass of oxygen in compound Z.

....

The mass of carbon present in compound Z is ....

....

The mass of hydrogen present in compound Z is ....

.... **(3 marks)**

(b) Use your answer to part (a) to calculate the empirical formula of compound Z.

....

....

.... **(3 marks)**

(c) The molar mass of compound Z is $90\,g\,mol^{-1}$.
Calculate its molecular formula.

....

.... **(2 marks)**

High-resolution mass spectra

Use these relative atomic masses when you answer these questions.

Element	Symbol	A_r
Hydrogen	H	1.0078
Carbon	C	12.0107
Oxygen	O	15.9994
Chlorine	Cl	35.4532

1 The mass to charge ratio, m/z, of the molecular ion, Cl_2^+, is:

☐ **A** 71.0

☐ **B** 17.7266

☐ **C** 35.4532

☐ **D** 70.9064 **(1 mark)**

2 Methanoic acid, HCOOH, and ethanol, CH_3CH_2OH, produce molecular ions with $m/z = 46.0$ in their low-resolution mass spectra. The charge on the ions is 1+, so the M_r is 46.0 for these compounds.

(a) Calculate the high-resolution M_r values of the molecular ions of:

(i) methanoic acid: ....

....

(ii) ethanol: ....

.... **(2 marks)**

(b) The molecular ion of methoxymethane, CH_3OCH_3, has the same m/z ratio as the molecular ion of ethanol. Describe the additional information found in a mass spectrum that would allow you to distinguish the two compounds.

....

.... **(2 marks)**

3 1-chloropropane, $CH_3CH_2CH_2Cl$, and ethanoyl chloride, CH_3COCl, produce molecular ions with the same m/z ratio in their low-resolution mass spectra.

(a) Calculate the high-resolution M_r values of the molecular ions of:

(i) 1-chloropropane: ....

....

(ii) ethanoyl chloride: ....

.... **(2 marks)**

(b) State why you cannot distinguish 1-chloropropane from 2-chloropropane using the m/z ratios of their molecular ions, even in a high-resolution mass spectrum.

.... **(1 mark)**

^{13}C NMR spectroscopy

This chart may help you answer these questions.

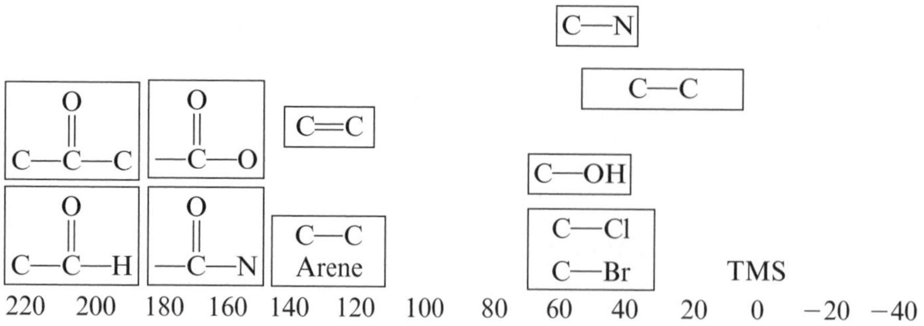

C—N

C—C

$$\begin{array}{cc} O & O \\ \| & \| \\ C-C-C & -C-O \end{array}$$ C=C

$$\begin{array}{cc} O & O \\ \| & \| \\ C-C-H & -C-N \end{array}$$ C—C Arene

C—OH

C—Cl
C—Br   TMS

| 220 | 200 | 180 | 160 | 140 | 120 | 100 | 80 | 60 | 40 | 20 | 0 | −20 | −40 |

Chemical shift relative to tetramethylsilane (TMS) /ppm

1 Predict the number of peaks present in the ^{13}C NMR spectrum of benzene, C_6H_6.

☐ **A** One

☐ **B** Two

☐ **C** Three

☐ **D** Six

> Draw the structure of benzene and count the number of different chemical environments.

(1 mark)

2 Propanal and propanone are functional group isomers of C_3H_6O. The table shows the chemical shifts, δ, for the peaks seen in the ^{13}C NMR spectrum of propanal, CH_3CH_2CHO.

Peak	δ /ppm
1	6.0
2	37.3
3	203.2

(a) (i) State the number of chemical environments present in propanal.

.... **(1 mark)**

(ii) Identify the chemical environment responsible for peak 3.

.... **(1 mark)**

(iii) Identify the chemical environments responsible for peaks 1 and 2.

.... **(1 mark)**

(b) Predict the number of peaks seen in the ^{13}C NMR spectrum of propanone, CH_3COCH_3, and explain your answer.

.... **(3 marks)**

> **Guided**

3 The diagram shows the skeletal formulae of two position isomers of cyclohexadiene.

Explain, using the diagrams, why the ^{13}C NMR spectrum of cyclohexa-1,3-diene shows 3 peaks but the spectrum of cyclohexa-1,4-diene only shows 2 peaks.

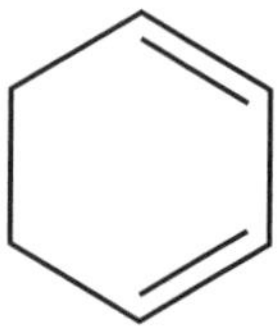

cyclohexa-1,3-diene

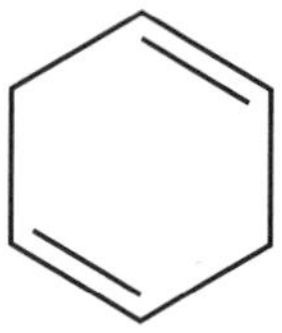

cyclohexa-1,4-diene

The number of chemical environments in cyclohexa-1,3-diene is

....

.... **(3 marks)**

Proton NMR spectroscopy

1   How many peaks would you expect to see in the low-resolution proton NMR spectrum of butanoic acid, $CH_3CH_2CH_2COOH$?

    ☐ **A**   Three

    ☐ **B**   Four

    ☐ **C**   Seven

    ☐ **D**   Eight

> Draw the displayed formula of butanoic acid and circle each chemical environment.

(1 mark)

2   Give three reasons why TMS is chosen as a reference standard in NMR spectroscopy.

....

....

.... **(3 marks)**

3   The solvents used in proton NMR spectroscopy must not produce their own peaks. State why each of the following solvents is suitable.

    (a)   tetrachloromethane, CCl_4.

.... **(1 mark)**

    (b)   deuterated trichloromethane, $CDCl_3$.

> Deuterium is ^2H.

.... **(1 mark)**

4   The diagram shows the low-resolution proton NMR spectrum for methanol, CH_3OH.

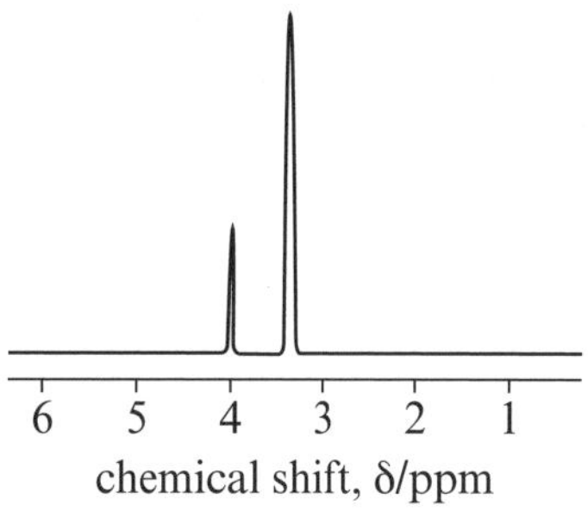

    (a)   State why there are two peaks in the spectrum.

.... **(1 mark)**

    (b)   Explain the difference in the area under each peak.

....

.... **(2 marks)**

5   The diagram shows the displayed formula of 2-bromobutane.

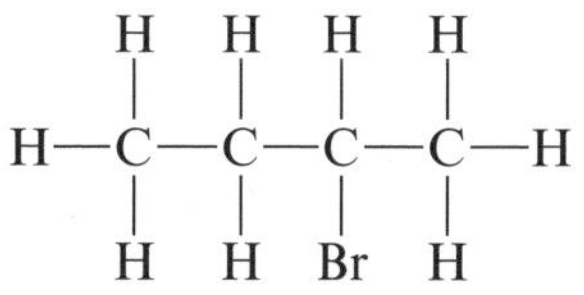

    (a)   Predict the number of peaks that would be present in its low-resolution proton NMR spectrum.

.... **(1 mark)**

    (b)   Predict the peak area ratios of the peaks in part (a), and identify each chemical environment.

....

.... **(2 marks)**

Splitting patterns

1 In a high-resolution proton NMR spectrum of methanol, CH_3OH, the peak due to the hydrogen atoms in the methyl group would be:

☐ **A** a singlet

☐ **B** a doublet

☐ **C** a triplet

☐ **D** a quartet

(1 mark)

2 This question is about the high-resolution proton NMR spectrum of 2-chlorobutane (the displayed formula is shown here).

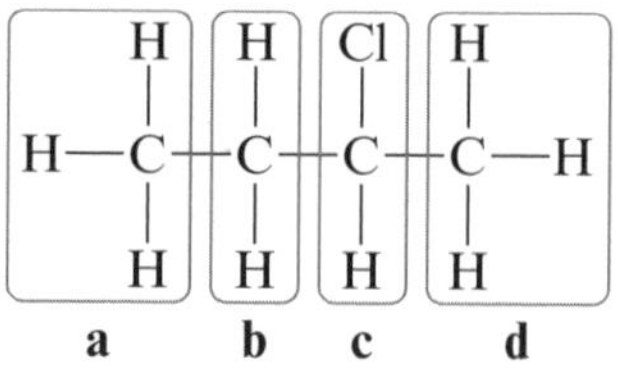

(a) Explain why the peak due to the hydrogen atoms in the methyl group labelled **a** is a triplet.

....

.... **(2 marks)**

(b) Explain why the peak due to the hydrogen atoms in the methyl group labelled **d** is a doublet.

....

.... **(2 marks)**

(c) Predict the splitting of the peak due to the hydrogen atoms in environment **b**, and explain your answer.

> The peak due to the H atoms in environment **c** is split into a sextet: 2 from **b**, 3 from **d**, and $(2 + 3) + 1 = 6$ using the $n + 1$ rule.

....

....

.... **(3 marks)**

3 Compound X has the molecular formula C_3H_8O. The table shows data from its high-resolution proton NMR spectrum.

Chemical shift, δ /ppm	Integration ratio	Splitting pattern
1.10	3	triplet
3.30	3	singlet
3.50	2	quartet

(a) Use the information to deduce the structure present in X responsible for the peak at:

(i) $\delta = 1.10$ ppm ....

(ii) $\delta = 3.30$ ppm ....

(iii) $\delta = 3.50$ ppm .... **(3 marks)**

(b) Give the displayed formula for X.

(1 mark)

Chemistry Practice Exam Paper

Advanced Subsidiary

Paper 1: Core Inorganic and Physical Chemistry

Time: 1 hour 30 minutes

You must have:
Data Booklet
Scientific calculator, ruler

Instructions

Use **black** ink or ball-point pen.
Answer **all** questions.

Information

- The total mark for this paper is 80.
- The marks for **each** question are shown in brackets
  – *use this as a guide as to how much time to spend on each question.*
- You may use a scientific calculator.
- For questions marked with an *, marks will be awarded for your ability to structure your answer logically showing the points that you make are related or follow on from each other where appropriate.
- A periodic table is printed on page 166 of this workbook.

Advice

- Read each question carefully before you start to answer it.
- Try to answer every question.
- Check your answers if you have time at the end.
- Show all your working in calculations and include units where appropriate.

Answer ALL questions.
Write your answers in the spaces provided.
Some questions must be answered with a cross in a box ☒.
If you change your mind about an answer, put a line through the box ☒
and then mark your new answer with a cross ☒

1 Potassium–40, ^{40}K, is the largest source of natural radioactivity in the human body.

  (a) Which of the following is the electronic configuration of the potassium ion, $^{40}K^+$?

    ☐ **A** $1s^2\ 2s^2\ 2p^6\ 3s^2\ 3p^6\ 4s^2$

    ☐ **B** $1s^2\ 2s^2\ 2p^6\ 3s^2\ 3p^6\ 4s^1$

    ☐ **C** $1s^2\ 2s^2\ 2p^6\ 3s^2\ 3p^6$

    ☐ **D** $1s^2\ 2s^2\ 2p^6\ 3s^2\ 3p^5$ **(1 mark)**

  (b) Potassium−40 is an isotope of potassium. Explain, in terms of sub-atomic particles, what is meant by the term **isotopes** of an element.

    (2 marks)

  (c) A sample of potassium contains 93.2% ^{39}K and 6.73% ^{41}K. The remainder of the sample is ^{40}K. Calculate the relative atomic mass, A_r, of potassium in this sample. Give your answer to an appropriate number of significant figures.

    (2 marks)

  (d) Table salt is sodium chloride, but 'low salt' alternatives contain a proportion of potassium chloride to reduce dietary intake of sodium ions. The recommended maximum daily intake of sodium chloride is 6 g.
Calculate the number of sodium ions, Na^+, in 6.00 g of sodium chloride, NaCl.

    (2 marks)
    (Total for question 1 = 7 marks)

2 The diagrams below represent the structures of beryllium chloride, sulfur dichloride and boron trichloride.

$$Cl—Be—Cl \qquad Cl\diagdown S \diagup Cl \qquad \underset{Cl \quad Cl}{\overset{Cl}{Cl—B}}$$

  (a) Explain, in terms of atomic structure, why chlorine is more electronegative than sulfur.

    (2 marks)

  (b) The bonds in all three molecules are polar, but SCl_2 is the only polar molecule. Explain why $BeCl_2$ and BCl_3 are non-polar molecules.

    (2 marks)

  (c) The diagram opposite represents the structure of a water molecule. Draw a diagram to show how a hydrogen bond can form between two water molecules. $H\diagdown \overset{O}{} \diagup H$

    (2 marks)
    (Total for question 2 = 6 marks)

3 The identity of the ions in a compound may be determined using simple analytical tests.

  (a) Which of the following is the flame test colour for caesium ions, Cs^+?

    ☐ **A** lilac

    ☐ **B** blue

    ☐ **C** red

    ☐ **D** red-violet   **(1 mark)**

  (b) For each pair of compounds, give a suitable reagent that could be added to distinguish between the two compounds, and describe your expected observations.

    (i) AgCl(s) and AgI(s)

    (3 marks)

    (ii) $HNO_3(aq)$ and HBr(aq)

    (3 marks)

  (c) The presence of sulfate ions can be determined using acidified barium chloride solution. Write the simplest ionic equation, including state symbols, for the reaction that occurs when acidified barium chloride solution is added to a solution containing sulfate ions.

    (2 marks)
    (Total for question 3 = 9 marks)

4 Chlorine is an element placed in Group 7 of the periodic table. It is a toxic substance used at water treatment plants and in swimming pools.

  (a) Which block in the periodic table contains the Group 7 elements?

    ☐ **A** *s* block

    ☐ **B** *p* block

    ☐ **C** *d* block

    ☐ **D** *f* block   **(1 mark)**

  (b) Calcium chloride is used to control the pH of swimming pool water. Which of the following could represent successive ionisation energies, in $kJ\ mol^{-1}$, for calcium?

☐ **A**	592	1146	4914	6475	8146
☐ **B**	421	3052	4414	5878	7977
☐ **C**	633	1236	2391	7090	8846
☐ **D**	1253	2298	3824	5159	6544

    (1 mark)

  (c) State why chlorine is added to drinking water.

    (1 mark)

  (d) Chlorine reacts with water, as shown by the following equation:

$$Cl_2(g) + H_2O(l) \rightleftharpoons HCl(aq) + HClO(aq)$$

    Which of the following shows the oxidation number of chlorine in each substance?

	Cl_2	HCl	HClO
☐ **A**	0	+1	−1
☐ **B**	+1	−1	0
☐ **C**	−1	+1	0
☐ **D**	0	−1	+1

    (1 mark)

  (e) Hydrogen peroxide, H_2O_2, has been used as an antiseptic. It decomposes in the presence of potassium iodide to form water and oxygen:

$$2H_2O_2(aq) \rightarrow 2H_2O(l) + O_2(g)$$

    Explain, in terms of oxidation numbers, why this is a disproportionation reaction.

    (2 marks)

  (f) A type of solder used to join the parts for car radiators contains lead and tin.
A 2.00 g sample of this solder was warmed with chlorine to form a mixture of insoluble lead(II) chloride, $PbCl_2$, and soluble tin(II) chloride, $SnCl_2$. The mixture was added to water, then filtered. The filtrate was dried, leaving a precipitate of mass 2.46 g.
Calculate the percentage by mass of lead in the solder, giving your answer to an appropriate number of significant figures.

    (4 marks)
    (Total for question 4 = 10 marks)

5 Group 2 carbonates decompose on heating to form Group 2 oxides and carbon dioxide:

$$MCO_3(s) \rightarrow MO(s) + CO_2(g)$$

A student investigated the relative thermal stability of Group 2 carbonates using the apparatus shown below.

boiling tube — delivery tube — group 2 carbonate — HEAT — limewater

  (a) State how the student could measure relative thermal stability in this investigation.

    (1 mark)

(b) Give **two** variables that the student could control to ensure that the carbonates are compared fairly.

(2 marks)

*(c) Explain the trend in thermal stability of Group 2 carbonates.

(4 marks)

(d) Lithium carbonate, $LiCO_3$, is a Group 1 carbonate that undergoes thermal decomposition. Write the balanced equation to show this reaction. State symbols are not required.

(1 mark)

(Total for question 5 = 8 marks)

6 Urea, $(NH_2)_2CO$, is widely used as a fertiliser to increase crop yields. It was the first organic compound to be synthesised artificially. Friedrich Wohler's original 1828 method involved the reaction between silver cyanate and ammonium chloride:

$AgNCO + NH_4Cl \rightarrow (NH_2)_2CO + AgCl$

(a) What is the shape of the ammonium ion, NH_4^+?

☐ **A** square planar

☐ **B** trigonal pyramidal

☐ **C** tetrahedral

☐ **D** octahedral **(1 mark)**

(b) Urea is also produced when carbonyl dichloride, $COCl_2$, reacts with ammonia, NH_3. Ammonium chloride is also produced in the reaction. Write the balanced equation to show this reaction. State symbols are not required.

(1 mark)

(c) Urea has a high nitrogen content compared to other common fertilisers. It contains 46.67% nitrogen, 6.67% hydrogen and 20.00% carbon by mass. The remainder is oxygen. Use this information to show that the empirical formula of urea is N_2H_4CO.

(3 marks)

(d) Complete the dot-and-cross diagram for the urea molecule, $(NH_2)_2CO$. Use dots (•) for the C and H electrons, and crosses (×) for the N and O electrons.

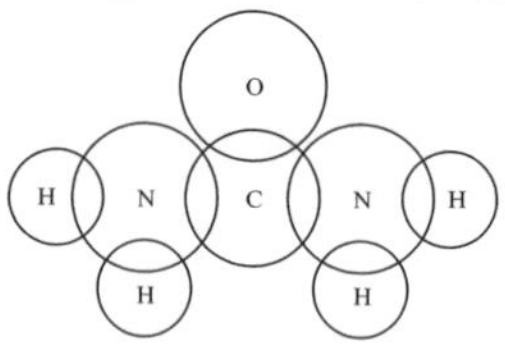

(2 marks)

(e) Deduce the shape of the urea molecule around the carbon atom, and justify your answer.

(3 marks)

(Total for question 6 = 10 marks)

7 Group 7 elements include fluorine, chlorine, bromine and iodine. They show trends in their physical and chemical properties.

(a) Which element in Group 7 has the highest electronegativity?

☐ **A** fluorine

☐ **B** chlorine

☐ **C** bromine

☐ **D** iodine **(1 mark)**

(b) Explain the trend in the physical state at room temperature in Group 7, going from fluorine to iodine.

(5 marks)

(c) Concentrated sulfuric acid, H_2SO_4, is added to solid potassium iodide. A mixture of products form, including hydrogen sulfide, H_2S, and iodine, I_2. The production of iodine from iodide ions is shown by this half-equation:

$2I^- \rightarrow I_2 + 2e^-$

State symbols are not required in your answers to parts (c)(i) and (ii).

(i) Write a half-equation to show the formation of H_2S from H_2SO_4.

(1 mark)

(ii) Write an overall equation for the reaction of I^- ions with H_2SO_4.

(1 mark)

(iii) Deduce the role of I^- ions in the reaction shown in part (c)(ii).

(1 mark)

(d) An equal volume of concentrated sulfuric acid is added to samples of solid potassium chloride and solid potassium iodide, and the reactions observed.

(i) Steamy fumes are produced in the reaction with potassium chloride. Name the gaseous product responsible.

(1 mark)

(ii) Steamy fumes are produced in the reaction with potassium bromide, but another reaction also takes place. Give an observation you would expect to see in this other reaction. Write balanced equations to show how this reaction occurs. State symbols are not required.

(3 marks)

(iii) State, in terms of reducing ability, why different products are formed in the reactions between solid halides and concentrated sulfuric acid.

(1 mark)

(Total for question 7 = 14 marks)

8 Sulfamic acid, NH_2SO_3H, is a white solid at room temperature. It is an ingredient in limescale removers, substances that react with limescale in kettles and steam irons in hard water areas. A sample of sulfamic acid was analysed to check its purity as part of the quality control procedures in a factory making a limescale remover. This is the method used.

1. A plastic weighing boat was placed on a 1 d.p. balance and the balance set to zero.
2. The sample of sulfamic acid was added to the weighing boat and the reading recorded.
3. The sample was transferred to a beaker and dissolved in deionised water.
4. The solution from step 3 was transferred to a 250 cm^3 volumetric flask and made up to the mark with deionised water.
5. A burette was filled with the sulfamic acid solution from step 4.
6. 25.00 cm^3 of 0.180 $mol\,dm^{-3}$ sodium hydroxide solution was transferred to a conical flask using a volumetric pipette, and titrated with the sulfamic acid solution. This was repeated several times.

(a) The results from the titrations are shown in the table below.

	Run 1	Run 2	Run 3	Run 4	Run 5
Final burette reading /cm^3	23.75	46.95	23.70	46.85	23.35
Initial burette reading /cm^3	0.00	23.70	0.20	23.70	0.15
Titre /cm^3					
Concordant titres (✓)					

Complete the table, and use the information to calculate the mean titre in cm^3.

(2 marks)

(b) The equation for the reaction of sulfamic acid with sodium hydroxide is:

$NH_2SO_3H(aq) + NaOH(aq) \rightarrow NH_2SO_3Na(aq) + H_2O(l)$

(i) Use the mean titre calculated in part (a) to calculate the concentration of sulfamic acid in $mol\,dm^{-3}$.

(2 marks)

(ii) Use your answer to part (b)(i) to calculate the concentration of sulfamic acid in $g\,dm^{-3}$.

(1 mark)

(iii) The reading on the balance at step 2 was 4.8 g. Calculate the percentage purity of the sulfamic acid sample.

(2 marks)

(c) Several measurements were made during the analysis. The table below shows the uncertainties for each measurement.

Measurement	Uncertainty	Error /%
Each balance reading	±0.05 g	
Volumetric flask volume	±0.5 cm^3	0.200
Volumetric pipette volume	±0.04 cm^3	
Each burette reading	±0.05 cm^3	0.413

(i) Complete the table (use these lines for your calculations).

(1 mark)

(ii) Estimate the total percentage error by adding together the four percentage errors in the table.

(1 mark)

(iii) The supplier of the sulfamic acid claims that is 99.0% pure. Use your answer to parts (b)(iii) and (c)(ii) to explain whether this claim is correct.

(1 mark)

(d) Give three sources of error in the method used for the analysis. For each source of error, suggest an appropriate improvement.

....

....

....

....

....

....

.... **(6 marks)**

(Total for question 8 = 16 marks)

TOTAL FOR PAPER = 80 MARKS

Chemistry Practice Exam Paper

Advanced Subsidiary
Paper 2: Core Organic and Physical Chemistry
Time: 1 hour 30 minutes

You must have:
Data Booklet
Scientific calculator, ruler

Instructions

Use **black** ink or ball-point pen.
Answer **all** questions.

Information

- The total mark for this paper is 80.
- The marks for **each** question are shown in brackets
  – *use this as a guide as to how much time to spend on each question.*
- You may use a scientific calculator.
- For questions marked with an *, marks will be awarded for your ability to structure your answer logically showing the points that you make are related or follow on from each other where appropriate.
- A periodic table is printed on page 166 of this workbook.

Advice

- Read each question carefully before you start to answer it.
- Try to answer every question.
- Check your answers if you have time at the end.
- Show all your working in calculations and include units where appropriate.

Answer ALL questions.
Write your answers in the spaces provided.
Some questions must be answered with a cross in a box ☒.
If you change your mind about an answer, put a line through the box ☒
and then mark your new answer with a cross ☒

1 Propane is an alkane used as a fuel in camping gas cylinders.

(a) Which value for x is required to balance this equation for the complete combustion of propane?

$$C_3H_8 + xO_2 \rightarrow 3CO_2 + 4H_2O$$

☐ **A** 3.5

☐ **B** 5

☐ **C** 7

☐ **D** 10

(1 mark)

(b) A student carries out an experiment to determine the relative formula mass of the gas in a propane cylinder. The student measures the mass of the cylinder and its contents, and opens the valve to collect some gas in an inverted measuring cylinder of water. The student then measures the mass of the cylinder and its contents again.

The table shows the results obtained by the student.

Mass of cylinder at start /g	Mass of cylinder at end /g	Volume of gas collected /cm³	Temperature /K	Pressure /Pa
2436.98	2436.52	238	293	101325

(i) Use the information in the table to calculate the mass of gas introduced to the measuring cylinder.

.... **(1 mark)**

(ii) Use the ideal gas equation to calculate the amount, in mol, of gas introduced into the measuring cylinder.

....

.... **(3 marks)**

(iii) Use your answers to part (b)(i) and (ii) to calculate the molar mass of the gas, giving your answer to one decimal place.

....

.... **(2 marks)**

(c) Butane, C_4H_{10}, is another alkane used as a fuel in camping gas cylinders.

(i) Draw the displayed formula of butane.

(1 mark)

(ii) State the empirical formula of butane.

.... **(1 mark)**

(Total for question 1 = 9 marks)

2 (a) In an exothermic reaction in aqueous solution, which of the following is correct?

	Sign of enthalpy change	Temperature
☐ **A**	positive	increases
☐ **B**	positive	decreases
☐ **C**	negative	increases
☐ **D**	negative	decreases

(1 mark)

(b) State the physical property that should be kept constant when measuring an enthalpy change.

.... **(1 mark)**

(c) Which equation represents the reaction for which the enthalpy change is the standard enthalpy change of formation, ΔH_f°, of sodium carbonate, Na_2CO_3?

☐ **A** $4Na(s) + 2C(s) + 3O_2(g) \rightarrow 2Na_2CO_3(s)$

☐ **B** $2Na(s) + C(s) + 1\frac{1}{2}O_2(g) \rightarrow Na_2CO_3(s)$

☐ **C** $2Na(s) + C(s) + 3O(g) \rightarrow Na_2CO_3(s)$

☐ **D** $2Na(g) + C(g) + 3O(g) \rightarrow Na_2CO_3(s)$

(1 mark)

(d) Define the term **standard enthalpy change of neutralisation**. In your answer, make it clear what **standard** means in this context.

....

....

....

.... **(3 marks)**

(Total for question 2 = 6 marks)

3 One of the stages in the manufacture of sulfuric acid involves the reaction of sulfur dioxide with oxygen:

$$2SO_2(g) + O_2(g) \rightleftharpoons 2SO_3(g) \qquad \Delta_r H^\circ = -197\,kJ\,mol^{-1}$$

*(a) A sulfuric acid manufacturer carries out this process at a temperature of 725 K and a pressure of 2 atmospheres. Evaluate the effect on the yield and rate of reaction of reducing the temperature to 500 K.

....

....

....

....

....

....

.... **(6 marks)**

(b) A vanadium(V) oxide catalyst is used by the manufacturer. The diagram below shows the reaction profile for the catalysed reaction.

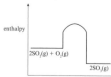

(i) On the diagram above, draw the reaction profile for the uncatalysed reaction. **(1 mark)**

(ii) Add arrows and labels to the diagram above to show, for the catalysed reaction:
- the enthalpy change of reaction, $\Delta_r H$
- the activation energy, E_a

(2 marks)

(c) Write the expression for the equilibrium constant, K_c, for the reaction between sulfur dioxide and oxygen.

.... **(1 mark)**

(Total for question 3 = 10 marks)

4 Nitrogen dioxide gas decomposes to form nitrogen monoxide and oxygen:

$$2NO_2(g) \rightarrow 2NO(g) + O_2(g) \qquad \Delta_r H^\circ = +114\,kJ\,mol^{-1}$$

(a) The diagram below shows the Maxwell–Boltzmann curve for the distribution of molecular energies for a fixed amount of nitrogen dioxide at a certain temperature. The activation energy in the absence of a catalyst is shown by the vertical line labelled E_a.

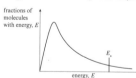

(i) Draw a vertical line on the graph to represent the most probable energy of the molecules. Label this line **A**. **(1 mark)**

(ii) On the same axes, sketch the distribution for the same sample of gas but at a **higher** temperature. **(2 marks)**

(b) The reaction is catalysed by nickel oxide.

(i) Draw a vertical line on your diagram to represent the activation energy of the catalysed reaction. Label this line **B**. **(1 mark)**

(ii) Explain, with the help of your diagram, why the use of a catalyst increases the rate of decomposition of nitrogen dioxide.

....

....

....

.... **(3 marks)**

(Total for question 4 = 7 marks)

5 A student carried out an experiment to determine the enthalpy change when ethanol was burned. Their method was:

1. Add 100.0 g of water to a copper can.
2. Measure and record the mass of spirit burner containing ethanol.
3. Measure and record the temperature of the water.
4. Light the wick, and use the flame to heat the water while stirring the water.
5. Put out the flame, then measure and record the mass of the spirit burner and the maximum temperature reached by the water.

The results from the experiment are shown below.

Mass of water	200.0 g
Initial mass of spirit burner	243.52 g
Final mass of spirit burner	242.64 g
Initial temperature of water	19.8 °C
Maximum temperature of water	34.3 °C

(a) Calculate:

(i) the mass of ethanol burned

.... **(1 mark)**

(ii) the change in temperature of the water

.... **(1 mark)**

(b) Calculate the enthalpy change of combustion of ethanol, in kJ mol⁻¹, giving your answer to three significant figures. (The specific heat capacity of water is $4.18\,J\,K^{-1}\,g^{-1}$)

....

....

....

.... **(4 marks)**

(c) A data book gives the standard enthalpy change of combustion of ethanol as $-1367\,kJ\,mol^{-1}$. Give two reasons why the student's value may differ from this value.

....

....

.... **(2 marks)**

(d) The reaction of ethanol with oxygen is shown by this equation:

$$H-\overset{\overset{\displaystyle H}{|}}{\underset{\underset{\displaystyle H}{|}}{C}}-\overset{\overset{\displaystyle H}{|}}{\underset{\underset{\displaystyle H}{|}}{C}}-O-H \; + \; 3\left[O=O\right] \longrightarrow 2\left[O=C=O\right] + 3\left[\begin{smallmatrix}H & & H\\ & O & \end{smallmatrix}\right]$$

The table below shows the mean bond enthalpies for bonds contained in these molecules.

Bond	Mean bond enthalpy /kJ mol⁻¹
C–C	+347
C–H	+413
C–O	+358
O–H	+464
O=O	+498
C=O	+805

(d) (i) Use this data to calculate the standard enthalpy change of combustion for ethanol in kJ mol⁻¹.

....

....

....

....

....

.... **(3 marks)**

(ii) Give two reasons why, other than error, the value calculated in part (d)(i) may differ from the data book value given in part (c).

.... **(2 marks)**

(Total for question 5 = 13 marks)

6 Ethane, C_2H_6, reacts with chlorine, Cl_2, at room temperature and pressure to produce 1-chloroethane, CH_3CH_2Cl:

$$C_2H_6 + Cl_2 \rightarrow CH_3CH_2Cl + HCl$$

(a) What is the mechanism for this reaction?

☐ **A** Electrophilic addition

☐ **B** Electrophilic substitution

☐ **C** Free radical addition

☐ **D** Free radical substitution **(1 mark)**

(b) State an essential condition for this reaction.

.... **(1 mark)**

(c) Write equations to show:

(i) the initiation step for this reaction.

.... **(1 mark)**

(ii) two propagation steps in this reaction.

....

.... **(2 marks)**

(iii) a termination step to produce 1-chloroethane.

.... **(1 mark)**

(d) Butane is also produced in the reaction.

(i) Explain, with the help of an equation, why this happens.

....

.... **(2 marks)**

(ii) Suggest a method to separate 1-chloroethane from the reaction mixture.

.... **(1 mark)**

(Total for question 6 = 9 marks)

7 Pent-1-ene is an unsaturated hydrocarbon with the molecular formula C_5H_{10}.

(a) Pent-1-ene reacts with bromine.

(i) Write an equation for the reaction and name the product formed.

....

.... **(2 marks)**

(ii) Describe the colour change you expect to see if a few drops of bromine are added to pent-1-ene.

.... **(1 mark)**

(b) Pent-2-ene is another isomer of C_5H_{10}.

(i) Pent-2-ene shows E/Z isomerism. Draw the displayed formulae of E-pent-2-ene and Z-pent-2-ene. Label each diagram.

(2 marks)

(ii) Explain why pent-2-ene shows E/Z isomerism but pent-1-ene does not.

....

.... **(2 marks)**

(c) Pent-1-ene and pent-2-ene are both unsaturated compounds.

(i) State the meaning of the term **unsaturated** in this context.

.... **(1 mark)**

(ii) There are isomers of C_5H_{10} that are saturated. Draw the skeletal formula for one of these isomers.

(1 mark)

(d) Pent-1-ene reacts with hydrogen bromide to form two products, 1-bromopentane and 2-bromopentane.

(i) Name the type of reaction mechanism involved.

.... **(1 mark)**

(ii) Draw the mechanism for the reaction that produces 2-bromopentane.

(3 marks)

(Total for question 7 = 13 marks)

8 This question is about isomeric alcohols with the molecular formula $C_4H_{10}O$.

(a) Complete the table to show the skeletal formulae of the three isomers W, X and Y.

Alcohol	Description	Skeletal formula
W	unbranched primary alcohol	
X	secondary alcohol	
Y	tertiary alcohol	

(3 marks)

(b) Describe how you could distinguish between these four isomers using potassium dichromate(VI) and Fehling's solution. In your answer, identify each alcohol by its letter (W, X or Y) and state the observations you expect to see with each alcohol or its product.

....

....

....

....

....

.... **(6 marks)**

(c) 2-methylpropan-1-ol is a fourth isomeric alcohol with the molecular formula $C_4H_{10}O$. It was analysed using mass spectrometry. The most prominent peak in the mass spectrum occurred at $m/z = 43$, with another peak at $m/z = 31$.

(i) The molecular ion was responsible for a peak at $m/z = 74$. State the charge on this ion.

....

(1 mark)

(ii) Give the structural formulae of the fragments responsible for the peaks at $m/z = 43$ and $m/z = 31$.

....

....

(2 marks)

(iii) State whether this isomer is a primary, secondary or tertiary alcohol.

....

(1 mark)

(Total for question 8 = 13 marks)

TOTAL FOR PAPER = 80 MARKS

Chemistry Practice Exam Paper

Advanced

Paper 1: Advanced Inorganic and Physical Chemistry

Time: 1 hour 45 minutes

You must have:
Data Booklet
Scientific calculator, ruler

Instructions
Use **black** ink or ball-point pen.
Answer **all** questions.

Information
- The total mark for this paper is 90.
- The marks for **each** question are shown in brackets
  − *use this as a guide as to how much time to spend on each question.*
- You may use a scientific calculator.
- For questions marked with an *, marks will be awarded for your ability to structure your answer logically showing the points that you make are related or follow on from each other where appropriate.
- A periodic table is printed on page 166 of this workbook.

Advice
- Read each question carefully before you start to answer it.
- Try to answer every question.
- Check your answers if you have time at the end.
- Show all your working in calculations and include units where appropriate.

Answer ALL questions.
Write your answers in the spaces provided.
Some questions must be answered with a cross in a box ☒.
If you change your mind about an answer, put a line through the box ☒
and then mark your new answer with a cross ☒

1 This question is about the bonding and structure of molecules.

(a) In covalent bonding there are strong electrostatic attractions between:

☐ **A** pairs of electrons

☐ **B** two atomic nuclei

☐ **C** instantaneous dipoles

☐ **D** nuclei and bonding pairs of electrons   **(1 mark)**

(b) Which compound has polar molecules?

☐ **A** carbon dioxide, CO_2

☐ **B** tetrachloromethane, CCl_4

☐ **C** ethanol, CH_3CH_2OH

☐ **D** ethene, $CH_2=CH_2$ **(1 mark)**

(c) Which of the following covalent bonds is the longest?

☐ **A** F–F

☐ **B** Cl–Cl

☐ **C** Br–Br

☐ **D** I–I   **(1 mark)**

(d) The diagram below shows the dot-and-cross diagram for phosphine, PH_3, in the gas state.

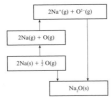

(i) Predict the shape of this molecule using the electron pair repulsion theory.

....
(1 mark)

(ii) Predict a value for the H–P–H bond angle, and explain your answer.

....
....
....
(3 marks)

(Total for question 1 = 7 marks)

2 This question is about energy changes involved in the formation of ionic compounds.

(a) What is the order of decreasing first ionisation energy for the elements chlorine, argon and potassium?

☐ **A** argon > chlorine > potassium

☐ **B** chlorine > potassium > argon

☐ **C** potassium > argon > chlorine

☐ **D** argon > potassium > chlorine   **(1 mark)**

(b) The third ionisation energy of aluminium has a magnitude of 2745 kJ mol⁻¹.
Which of the following represents the third ionisation energy of aluminium?

☐ **A** $Al(g) \rightarrow Al^{3+}(g) + 3e^-$ $\Delta H^\circ = -2745 \text{ kJ mol}^{-1}$

☐ **B** $Al^+(g) \rightarrow Al^{3+}(g) + 2e^-$ $\Delta H^\circ = +2745 \text{ kJ mol}^{-1}$

☐ **C** $Al^{2+}(g) \rightarrow Al^{3+}(g) + e^-$ $\Delta H^\circ = +2745 \text{ kJ mol}^{-1}$

☐ **D** $Al^{2+}(g) \rightarrow Al^{3+}(g) + e^-$ $\Delta H^\circ = -2745 \text{ kJ mol}^{-1}$ **(1 mark)**

(c) Define the term **first ionisation energy**.

....
....
....
(3 marks)

(d) The diagram below shows a Born–Haber cycle for sodium oxide, Na_2O.

$$2Na^+(g) + O^{2-}(g)$$
$$2Na(g) + O(g)$$
$$2Na(s) + \tfrac{1}{2}O(g)$$
$$Na_2O(s)$$

	kJ mol⁻¹
Enthalpy of atomisation of sodium, $Na(s) \rightarrow Na(g)$	107
Enthalpy of atomisation of oxygen, $\frac{1}{2}O_2(g) \rightarrow O(g)$	249
First ionisation energy of $Na(g)$	496
First electron affinity of $O(g)$	−141
Lattice energy of $Na_2O(s)$	−2478
Enthalpy of formation of $Na_2O(s)$	−414

The change represented by the second electron affinity of oxygen is shown by this equation:

$$O^-(g) + e^- \rightarrow O^{2-}(g)$$

Use the Born−Haber cycle and the data in the table to calculate the second electron affinity for oxygen ($E_{aff,2}$) in kJ mol⁻¹

....
....
....
....
....
....
....
....
(3 marks)

(Total for question 2 = 8 marks)

3 Manganese and vanadium are transition metals that form different ions.

(a) Complete the electronic configurations for the manganese atom and the manganese(II) ion.

(i) Mn $1s^2\,2s^2\,2p^6\,3s^2$ ....   **(1 mark)**

(ii) Mn^{2+} $1s^2\,2s^2\,2p^6\,3s^2$ ....   **(1 mark)**

(b) The table below shows the standard electrode (redox) potentials, E°, for some half-cell reactions.

Half-cell reaction	E°/V
$V^{2+}(aq) + 2e^- \rightleftharpoons V(s)$	−1.18
$V^{3+}(aq) + e^- \rightleftharpoons V^{2+}(aq)$	−0.26
$VO^{2+}(aq) + 2H^+(aq) + e^- \rightleftharpoons V^{3+}(aq) + H_2O(l)$	+0.34
$VO_2^+(aq) + 2H^+(aq) + e^- \rightleftharpoons VO^{2+}(aq) + H_2O(l)$	+1.00
$MnO_4^-(aq) + 8H^+(aq) + 5e^- \rightleftharpoons Mn^{2+}(aq) + 4H_2O(l)$	+1.51

Excess $KMnO_4(aq)$ was added to a solution containing $V^{2+}(aq)$ ions.

(i) Use the data from the table above to predict the final vanadium species present at the end of the reaction. Justify your answer.

....
....
....
(3 marks)

(ii) Write a half-equation to show the formation of the final vanadium species.

....
(1 mark)

(iii) Write an equation to show the overall reaction between $MnO_4^-(aq)$ and $V^{2+}(aq)$ to form the final vanadium species.

....
....
(1 mark)

(Total for question 3 = 7 marks)

4 This question is about complex ions and ligands.

(a) 1,2-diaminoethane, $H_2NCH_2CH_2NH_2$, is a bidentate ligand. Explain the meaning of the term **bidentate** in this context.

....
....
....
(2 marks)

(b) Hexaaquacopper(II) ions, $[Cu(H_2O)_6]^{2+}$, are present in copper(II) sulfate solution.
They react with an excess of 1,2-diaminoethane to form a dark blue solution.
This contains the complex ion, $[Cu(en)_x(H_2O)_y]^{2+}$, where en represents a molecule of 1,2-diaminoethane. The numbers represented by x and y are integers in the range 0 to 6.
An experiment was carried out to determine the formula of this complex ion.
This is the method used.
1. Eight test tubes were set up, each containing a different mixture of 0.050 mol dm⁻³ copper(II) sulfate solution and 0.100 mol dm⁻³ 1,2-diaminoethane solution.
2. The mixtures were filtered separately to remove any precipitate formed.
3. The absorbance of the filtrates, containing the $[Cu(en)_x(H_2O)_y]^{2+}$ ion, were measured using a colorimeter.

The results from the measurements are shown in the table below.

	Test tube number							
	1	2	3	4	5	6	7	8
Volume of 0.050 mol dm⁻³ CuSO₄(aq) /cm³	0.0	2.0	3.0	4.0	6.0	8.0	10.0	12.0
Volume of 0.100 mol dm⁻³ H₂NCH₂CH₂NH₂(aq) /cm³	10.0	10.0	9.0	8.0	6.0	4.0	2.0	0.0
Absorbance	0.00	0.45	0.67	0.90	0.92	0.64	0.36	0.08

(i) Plot a graph of absorbance against volume of CuSO₄(aq) on the grid.
Draw a straight line of best fit through the first four points, and another straight line of best fit through the remaining four points.
Extend both lines so that they cross.

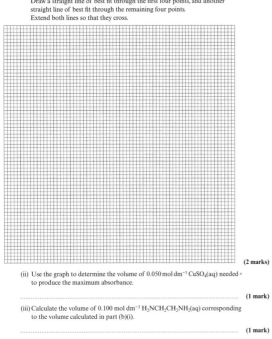

(2 marks)

(ii) Use the graph to determine the volume of 0.050 mol dm⁻³ CuSO₄(aq) needed to produce the maximum absorbance.

.... **(1 mark)**

(iii) Calculate the volume of 0.100 mol dm⁻³ H₂NCH₂CH₂NH₂(aq) corresponding to the volume calculated in part (b)(i).

.... **(1 mark)**

(c) Calculate the amounts, in mol, of CuSO₄(aq) and H₂NCH₂CH₂NH₂(aq) in the volumes calculated in parts (b)(ii) and (iii).

....
....
....
.... **(2 marks)**

(d) Use your answers to part (c) to:

(i) Calculate the ratio of H₂NCH₂CH₂NH₂ ligands to Cu²⁺ ions.

.... **(1 mark)**

(ii) Determine the formula for the complex ion formed.

.... **(1 mark)**

(Total for question 4 = 10 marks)

5 Chlorine and iodine are elements in Group 7 of the periodic table. They can undergo redox reactions.

(a) Chlorine reacts with cold, dilute sodium hydroxide solution:

Cl₂(aq) + 2NaOH(aq) → NaCl(aq) + NaClO(aq) + H₂O(l)

(i) Give a household use of the product NaClO(aq).

.... **(1 mark)**

(ii) When it is heated above 75 °C, NaClO(aq) forms NaClO₃(aq), which is used as a weedkiller:

3NaClO(aq) → 2NaCl(aq) + NaClO₃(aq)

Explain, using oxidation numbers, why this is a disproportionation reaction.

....
....
.... **(3 marks)**

(b) Hydrogen chloride, a colourless gas, is produced when sodium chloride reacts with concentrated sulfuric acid.

(i) Explain whether or not this is a redox reaction.

....
.... **(2 marks)**

(ii) Explain why hydrogen chloride forms misty fumes in moist air.

....
.... **(2 marks)**

(c) A mixture of products form when concentrated sulfuric acid is added to potassium iodide. These including sulfur dioxide and iodine, I₂.
State symbols are not required in your answers to parts (c)(i) and (ii).

(i) Write a half-equation to show the formation of sulfur dioxide from sulfuric acid.

.... **(1 mark)**

(ii) Write an overall equation for the reaction of iodide ions with concentrated sulfuric acid to produce sulfur dioxide.

.... **(1 mark)**

(iii) Deduce the role of iodide ions in the reaction shown in part (c)(ii).

.... **(1 mark)**

(Total for question 5 = 11 marks)

6 An average adult releases 345 dm³ of carbon dioxide each day due to respiration. Spacecraft need carbon dioxide 'scrubbers' to remove the carbon dioxide released by astronauts. Lithium hydroxide, LiOH, was used in the Apollo spacecraft which sent people to the Moon and back in the last century. Lithium hydroxide reacts with carbon dioxide to produce lithium carbonate and water vapour:

2LiOH(s) + CO₂(g) → Li₂CO₃(s) + H₂O(g)

(a) Calculate the amount, in mol, of carbon dioxide released by an average adult each day. (Molar volume of an ideal gas at r.t.p. = 24.0 dm³ mol⁻¹)

.... **(1 mark)**

(b) Use your answer to part (a) to calculate the mass of lithium hydroxide needed to absorb the carbon dioxide produced by a crew of three astronauts on an 8-day mission. Give your answer in kg to three significant figures.

....
....
....
.... **(4 marks)**

(c) Modern carbon dioxide scrubbers may use lithium peroxide, Li₂O₂, instead:

2Li₂O₂(s) + 2CO₂(g) → 2Li₂CO₃(s) + O₂(g)

Calculate the mass of lithium peroxide needed to absorb the same amount of carbon dioxide released during the mission described in part (b).

....
....
.... **(2 marks)**

(d) Suggest two advantages of lithium peroxide, rather than lithium hydroxide, for use in carbon dioxide scrubbers in spacecraft.

....
.... **(2 marks)**

(Total for question 6 = 9 marks)

---

7 Hydrogen can be produced on an industrial scale by the reaction of methane with steam:

CH₄(g) + 2H₂O(g) ⇌ CO₂(g) + 4H₂(g)   ΔH⦵ = +165 kJ mol⁻¹

(a) Which of the changes below would affect the value of the equilibrium constant, K_c, **and** increase the proportion of hydrogen present in an equilibrium mixture of the gases?

☐ **A** reducing the pressure

☐ **B** adding a catalyst

☐ **C** increasing the temperature

☐ **D** decreasing the temperature **(1 mark)**

(b) Write an expression for the equilibrium constant, K_c, for this reaction.

....
.... **(1 mark)**

(c) 0.80 mol of methane and 1.6 mol of steam were sealed in a 2.5 dm³ flask, and heated in the presence of a catalyst. At equilibrium, the flask was found to contain 0.20 mol of carbon dioxide.

(i) Calculate the equilibrium amounts, in mol, of the other three gases.

....
....
.... **(3 marks)**

(ii) Use your answers to part (c)(i) to calculate the equilibrium concentrations of all four gases.

....
....
....
.... **(1 mark)**

(iii) Use your answers to parts (b) and (c)(ii) to calculate the value of K_c for this reaction. State the units.

....
....
.... **(3 marks)**

(Total for question 7 = 9 marks)

8 This question is about acids and bases.

(a) Ethanoic acid reacts with water:

CH₃COOH(aq) + H₂O(l) ⇌ CH₃COO⁻(aq) + H₃O⁺(aq)

The Brønsted−Lowry bases in this equilibrium are:

☐ **A** H₂O and H₃O⁺

☐ **B** H₂O and CH₃COO⁻

☐ **C** CH₃COO⁻ and H₃O⁺

☐ **D** CH₃COOH and H₃O⁺ **(1 mark)**

(b) State why ethanoic acid is classified as a **weak acid**.

....
(1 mark)

(c) A school experiment requires hydrochloric acid with a pH of 0.30. Calculate the concentration of hydrochloric acid needed.

....

....
(1 mark)

(d) A buffer solution is prepared by mixing $125\,cm^3$ of $0.250\,mol\,dm^{-3}$ ethanoic acid solution with $75\,cm^3$ of $0.100\,mol\,dm^{-3}$ sodium ethanoate solution, $CH_3COONa(aq)$.

(i) Explain what is meant by the term **buffer solution**.

....

....
(2 marks)

(ii) Write an equation to show what happens when a small amount of hydrochloric acid is added to this buffer solution.

....
(1 mark)

(iii) Calculate the pH of this buffer solution. (K_a for $CH_3COOH = 1.74 \times 10^{-5}\,mol\,dm^{-3}$)

....

....

....

....

....

....

....
(4 marks)

(Total for question 8 = 10 marks)

9 (a) Platin, $Pt[Cl_2(NH_3)_2]$, exists as two geometrical isomers. These are *cis*-platin (used as an anticancer drug) and *trans*-platin.

(i) Platin has geometrical isomers because:

☐ **A** it is tetrahedral in shape

☐ **B** it is octahedral in shape

☐ **C** it is square planar in shape

☐ **D** it contains a double bond.
(1 mark)

(ii) The oxidation number of Pt in *trans*-platin is:

☐ **A** +2

☐ **B** +1

☐ **C** −1

☐ **D** −2
(1 mark)

(b) Cobalt(II) sulfate dissolves in water to form a pink solution.

(i) Give the formula of the cobalt complex present in the solution.

....
(1 mark)

(ii) State the shape of the complex in part (i).

....
(1 mark)

(iii) A new, straw-coloured, complex forms when an excess of concentrated ammonia solution is added to cobalt(II) sulfate solution. Write an equation for the formation of this complex, and state the type of reaction that occurs.

....

....
(2 marks)

(Total for question 9 = 6 marks)

10 This question is about compounds of Group 1 and 2 elements.

(a) Which of the following shows the correct trends in solubility of Group 2 hydroxides and sulfates as Group 2 is descended?

		Hydroxides	Sulfates
☐	**A**	decreases	increases
☐	**B**	decreases	decreases
☐	**C**	increases	increases
☐	**D**	increases	decreases

(1 mark)

(b) Explain the trend in thermal stability of Group 2 nitrates.

....

....

....

....
(3 marks)

(c) Potassium chloride is soluble in water. The enthalpy level diagram for the processes occurring when potassium chloride is dissolved in water is shown below.

Use the data in the table below to calculate the enthalpy change of solution, $\Delta H_{solution}$, of $KCl(s)$ in $kJ\,mol^{-1}$.

	$kJ\,mol^{-1}$
Lattice energy of $KCl(s)$	−711
Enthalpy change of hydration of $K^+(g)$	−351
Enthalpy change of hydration of $Cl^-(g)$	−364

....

....

....
(2 marks)

(d) Calcium carbonate, $CaCO_3$, decomposes to form calcium oxide and carbon dioxide:

$$CaCO_3(s) \rightarrow CaO(s) + CO_2(g)$$

The table below gives some information about these substances.

Substance	Standard enthalpy change of formation, $\Delta_f H^\circ$/$kJ\,mol^{-1}$	Standard entropy, S°/$J\,K^{-1}\,mol^{-1}$
$CaCO_3(s)$	−1207	93
$CaO(s)$	−635	40
$CO_2(g)$	−394	214

(i) Calculate the standard enthalpy change of reaction, $\Delta_r H^\circ$.

....

....

....
(1 mark)

(ii) Calculate the standard entropy change of the system, ΔS°_{system}.

....

....

....
(1 mark)

(iii) State, in terms of particles, the meaning of the sign of your answer to part (d)(ii).

....
(1 mark)

(iv) Use the equation $\Delta G = \Delta H - T\Delta S^\circ_{system}$ to determine whether the reaction is feasible under standard conditions.

....

....

....
(2 marks)

(v) Calculate the minimum temperature at which the reaction becomes feasible.

....

....

....
(2 marks)

(Total for question 10 = 13 marks)

TOTAL FOR PAPER = 90 MARKS

Chemistry Practice Exam Paper

Advanced
Paper 2: Advanced Organic and Physical Chemistry
Time: 1 hour 45 minutes

You must have:
Data Booklet
Scientific calculator, ruler

Instructions
Use **black** ink or ball-point pen.
Answer **all** questions.

Information
- The total mark for this paper is 90.
- The marks for **each** question are shown in brackets
  – use this as a guide as to how much time to spend on each question.
- You may use a scientific calculator.
- For questions marked with an *, marks will be awarded for your ability to structure your answer logically showing the points that you make are related or follow on from each other where appropriate.
- A periodic table is printed on page 166 of this workbook.

Advice
- Read each question carefully before you start to answer it.
- Try to answer every question.
- Check your answers if you have time at the end.
- Show all your working in calculations and include units where appropriate.

Answer ALL questions.
Write your answers in the spaces provided.
Some questions must be answered with a cross in a box ⊠.
If you change your mind about an answer, put a line through the box ⊠
and then mark your new answer with a cross ⊠

1 This question is about some hydrocarbons.

(a) What is the systematic (IUPAC) name for the hydrocarbon shown opposite?

   □ **A** 3-ethyl-3-methylpentane

   □ **B** 2-methyl-2-ethylpentane

   □ **C** 2,2,2-triethylethane

   □ **D** 2,2-diethylbutane **(1 mark)**

1

(b) How many isomers of C_6H_{14} are there?

   □ **A** three

   □ **B** four

   □ **C** five

   □ **D** six **(1 mark)**

(c) Alkenes contain carbon−carbon double bonds. Describe how the orbitals from carbon atoms interact to form these bonds. You may include a labelled diagram to support your answer.

....

....

....

(3 marks)

(Total for question 5 = 5 marks)

2 The skeletal formulae of four different organic compounds are shown below.

(a) Which of these compounds form a pale yellow precipitate when added to a mixture of iodine and sodium hydroxide solution?

   □ **A** Compound 1 only

   □ **B** Compounds 2 and 3

   □ **C** Compounds 1, 2 and 4

   □ **D** Compounds 2 and 4 **(1 mark)**

2

(b) Which of these compounds forms a silver mirror with Tollens' reagent but **not** a pale yellow precipitate when added to a mixture of iodine and sodium hydroxide solution?

   □ **A** Compound 1

   □ **B** Compound 2

   □ **C** Compound 3

   □ **D** Compound 4 **(1 mark)**

(c) Which of these compounds form an orange-yellow precipitate when added to 2,4-dinitrophenylhydrazine (Brady's reagent)?

   □ **A** Compounds 1 and 2

   □ **B** Compound 3 only

   □ **C** Compound 1, 2 and 4

   □ **D** Compounds 1 and 4 **(1 mark)**

(Total for question 2 = 3 marks)

3 This question is about amines.

(a) Amines can behave as Brønsted-Lowry bases. State the meaning of the term **base** in this context.

.... **(1 mark)**

(b) Which of these compounds is the weakest base?

   □ **A** NH_3

   □ **B** CH_3NH_2

   □ **C** $C_6H_5NH_2$

   □ **D** $C_6H_5CH_2NH_2$ **(1 mark)**

(c) Ethylamine, $CH_3CH_2NH_2$, reacts with acids to form salts. Give the formula of the salt formed when ethylamine reacts with:

(i) nitric acid, HNO_3

.... **(1 mark)**

(ii) sulfuric acid, H_2SO_4

.... **(1 mark)**

(d) Butylamine, $CH_3CH_2CH_2CH_2NH_2$ is soluble in water.

(i) Explain, in terms of intermolecular forces, why butylamine is soluble in water.

....

.... **(2 marks)**

(ii) Write an equation for the reaction that occurs when butylamine reacts with water.

.... **(1 mark)**

(Total for question 3 = 7 marks)

3

---

4 This question is about amino acids.

(a) The diagrams below show the structures of two naturally-occurring amino acids, glycine and alanine.

   glycine   alanine

   Give the systematic (IUPAC) names for these two amino acids.

....

.... **(2 marks)**

(b) Explain which of these amino acids has optical isomers.

....

.... **(2 marks)**

(c) Give the structural formula of glycine in:

(i) acidic solution

.... **(1 mark)**

(ii) alkaline solution

.... **(1 mark)**

*(d) The table below shows some information about alanine and two other organic compounds.

Compound	Molar mass /g mol⁻¹	Melting temperature /K
alanine	89.0	531
butanoic acid	88.0	268
hexane	86.0	178

Explain, in terms of intermolecular forces and using information from the table, the differences in melting temperatures of these compounds.

....

....

....

....

....

(6 marks)

(Total for question 4 = 12 marks)

5 This question is about alkenes.

(a) Alkenes undergo electrophilic addition reactions. An electrophile is defined as:

   □ **A** a negatively charged species

   □ **B** an electron-rich species

   □ **C** an electron pair acceptor

   □ **D** an electron pair donor **(1 mark)**

4

(b) Unsaturated vegetable oils can be hydrogenated for use in the manufacture of margarine. The oils are mixed with hydrogen in the presence of:

☐ **A** an iron catalyst at 250 °C

☐ **B** a vanadium(V) oxide catalyst at 200 °C

☐ **C** a nickel catalyst at 60 °C

☐ **D** an alumina catalyst at 180 °C   **(1 mark)**

(c) Two compounds have the molecular formula C_6H_{12}. Compound A decolourises bromine water but compound B does not. Suggest why this happens.

....

.... **(2 marks)**

(d) (i) Give the mechanism for the reaction of but-1-ene with hydrogen bromide, HBr, to form the minor product.

                                                                          (3 marks)

(ii) Name the major product, and explain why it forms in preference to the minor product.

....

....

.... **(3 marks)**

(Total for question 5 = 10 marks)

---

6 This question is about the kinetics of chemical reactions.

(a) The rate equation for a reaction involving three reactants A, B and C is:

$$\text{Rate} = k[A]^2[C]$$

Which statement is correct?

☐ **A** The rate doubles if the concentration of A is halved and the concentration of C is increased by a factor of 4.

☐ **B** The rate doubles if the concentration of A is doubled and the concentration of C is halved.

☐ **C** The rate doubles if the concentration of A is doubled and the concentration of C is kept constant.

☐ **D** The rate is directly proportional to the concentration of reactant B. **(1 mark)**

(b) Describe what is meant by the term **order of reaction**, in the context of a reactant in a reaction.

....

.... **(2 marks)**

(c) Propanone and iodine react together in acidic conditions to form 1-iodopropan-2-one and hydrogen iodide:

$$CH_3COCH_3 + I_2 \rightarrow CH_2ICOCH_3 + HI$$

The initial rate of reaction was measured in a series of experiments in dilute hydrochloric acid, and this rate equation was determined:

$$\text{rate} = k[CH_3COCH_3][H^+]$$

(i) State the overall order of the reaction.

.... **(1 mark)**

(ii) Explain the role of hydrogen ions in this reaction.

....

.... **(2 marks)**

(iii) Sketch a graph of concentration of iodine against time for this reaction.

                                                                          (1 mark)

(iv) Explain, in terms of collision theory, why increasing the concentration of propanone changes the rate of this reaction.

....

....

.... **(2 marks)**

(d) Hydrogen and nitrogen monoxide react together to form water vapour and nitrogen:

$$2H_2(g) + 2NO(g) \rightarrow 2H_2O(g) + N_2(g)$$

The initial rates method was used to investigate the orders of reaction with respect to hydrogen and nitrogen monoxide. The table shows the results obtained.

Run	Initial $[H_2(g)]$ /mol dm^{-3}	Initial $[NO(g)]$ /mol dm^{-3}	Initial rate /mol dm^{-3} s^{-1}
1	0.85×10^{-3}	2.6×10^{-3}	1.08×10^{-5}
2	0.85×10^{-3}	3.9×10^{-3}	2.43×10^{-5}
3	1.70×10^{-3}	5.2×10^{-3}	8.64×10^{-5}

(i) Determine the order of reaction with respect to nitrogen monoxide.

.... **(1 mark)**

(ii) Use your answer to part (d)(i) and information in the table to determine the order of reaction with respect to hydrogen, and justify your answer.

....

....

.... **(2 marks)**

(iii) Write the rate equation for the reaction.

.... **(1 mark)**

(iv) Calculate the value of the rate constant, k, and state its units.

....

.... **(2 marks)**

(e) The diagram below shows the Maxwell−Boltzmann curve for the distribution of molecular energies for a fixed amount of gas at a certain temperature. The activation energy is shown by the vertical line labelled E_a.

(i) On the graph, sketch the distribution for the same sample of gas but at a **higher** temperature. **(2 marks)**

(ii) Explain, with reference to the graphs, how a change in temperature affects the rate of reaction. **(2 marks)**

(Total for question 6 = 19 marks)

---

7 Phenylamine, $C_6H_5NH_2$, is used in the manufacture of polymers and dyes. It can be synthesised from benzene in two stages:

benzene   nitrobenzene   phenylamine

(a) The reaction mechanism for the nitration of benzene is electrophilic substitution.

(i) State the reagents needed to produce the electrophile, and write an equation to show how it forms.

....

.... **(3 marks)**

(ii) Give the reaction mechanism for the nitration of benzene to produce nitrobenzene.

                                                                          (3 marks)

(b) Nitrobenzene can be reduced to form phenylamine.

(i) State a suitable reducing agent for this reaction.

.... **(1 mark)**

(ii) Write an equation for the reduction of nitrobenzene to phenylamine, using [H] to represent the reducing agent.

.... **(1 mark)**

(c) Azobenzene, $C_6H_5N=NC_6H_5$, is an orange-red solid that can be synthesised from phenylamine:

(i) The mass spectrum of azobenzene has prominent peaks at m/z 182, 105 and 77. Name the ion responsible for the peak at m/z 182, and give the formulae of the fragments responsible for the peaks at m/z 105 and 77.

....

....

.... **(3 marks)**

(ii) Azobenzene has *cis-trans* stereoisomers. The diagram above shows *trans*-azobenzene. Sketch the structure of *cis*-azobenzene, and suggest why these isomers exist even though azobenzene is not an alkene.

....

.... **(3 marks)**

(Total for question 7 = 14 marks)

8 This question is about polymers.

(a) Poly(chloroethene) can be synthesised from chloroethene.

(i) Give the structural formula of chloroethene.

.... **(1 mark)**

(ii) Give the displayed formula of poly(chloroethene).

(1 mark)

(iii) Explain, without calculation, the atom economy for the production of poly(chloroethene) from chloroethene.

....

.... **(2 marks)**

(b) The repeating unit for a polyester is shown below.

$$\left[O - \underset{\underset{H}{|}}{\overset{\overset{H}{|}}{C}} - \underset{\underset{H}{|}}{\overset{\overset{H}{|}}{C}} - O - \overset{\overset{O}{\|}}{C} - \underset{\underset{H}{|}}{\overset{\overset{H}{|}}{C}} - \underset{\underset{H}{|}}{\overset{\overset{H}{|}}{C}} - \overset{\overset{O}{\|}}{C} \right]$$

(i) Give the systematic (IUPAC) name for the alcohol used to make this polymer.

.... **(1 mark)**

(ii) Name the type of polymerisation involved to produce this polymer.

.... **(1 mark)**

(iii) Polyesters are often used to make fibres for clothing. Explain why spilled sodium hydroxide solution can make holes in such clothing.

....

....

.... **(3 marks)**

(c) A polyamide can be formed by the reaction of butanedioic acid and hexane-1,6-diamine. Give the repeating unit for this polyamide.

(1 mark)

(Total for question 8 = 10 marks)

9 Polylactic acid is a biodegradable polymer used to make plastic cups. It is manufactured from lactic acid, which can be produced from ethanol in three steps:

$$H - \underset{\underset{H}{|}}{\overset{\overset{H}{|}}{C}} - \underset{\underset{H}{|}}{\overset{\overset{H}{|}}{C}} - OH \xrightarrow{\text{Step 1}} H - \underset{\underset{H}{|}}{\overset{\overset{H}{|}}{C}} - \overset{\overset{O}{\diagup\!\!\diagdown}}{C} \xrightarrow{\text{Step 2}} H - \underset{\underset{H}{|}}{\overset{\overset{H}{|}}{C}} - \underset{\underset{H}{|}}{\overset{\overset{OH}{|}}{C}} - CN \xrightarrow{\text{Step 3}} H - \underset{\underset{H}{|}}{\overset{\overset{H}{|}}{C}} - \underset{\underset{OH}{|}}{\overset{\overset{OH}{|}}{C}} - \overset{\overset{O}{\diagup\!\!\diagdown}}{C}_{OH}$$

(a) (i) State the type of reaction involved in step 1.

.... **(1 mark)**

(ii) Give the reagents and reaction conditions necessary to carry out the reaction in step 1.

....

....

.... **(3 marks)**

(b) (i) State the type of reaction mechanism involved in step 2.

.... **(1 mark)**

(ii) Give the reagents necessary to carry out the reaction in step 2.

.... **(2 marks)**

(c) Explain, with a displayed formula, why the hydroxynitrile shown in step 2 contains a chiral centre, but the product of the reaction is not optically active.

(3 marks)

(Total for question 9 = 10 marks)

TOTAL FOR PAPER = 90 MARKS

Chemistry Practice Exam Paper

Advanced

Paper 3: General and Practical Principles in Chemistry

Time: 2 hours 30 minutes

You must have:
Data Booklet
Scientific calculator, ruler

Instructions

Use **black** ink or ball-point pen.
Answer **all** questions.

Information

- The total mark for this paper is 120.
- The marks for **each** question are shown in brackets
  – *use this as a guide as to how much time to spend on each question.*
- You may use a scientific calculator.
- For questions marked with an *, marks will be awarded for your ability to structure your answer logically showing the points that you make are related or follow on from each other where appropriate.
- A periodic table is printed on page 166 of this workbook.

Advice

- Read each question carefully before you start to answer it.
- Try to answer every question.
- Check your answers if you have time at the end.
- Show all your working in calculations and include units where appropriate.

Answer ALL questions.
Write your answers in the spaces provided.
Some questions must be answered with a cross in a box ☒.
If you change your mind about an answer, put a line through the box ☒
and then mark your new answer with a cross ☒

1 Halide ions in solution may be detected using acidified silver nitrate solution and aqueous ammonia.

(a) (i) Give the formulae and colours of the precipitates formed by chloride, bromide and iodide ions with silver nitrate solution.

....

.... **(3 marks)**

(ii) State why the presence of fluoride ions cannot be detected using silver nitrate solution.

.... **(1 mark)**

(iii) Name the acid used to acidify the sample in these tests, and state why it is needed.

....

.... **(2 marks)**

(b) Dilute ammonia solution and concentrated ammonia solution are used to confirm the identity of the halide ion present in a compound. Describe how these solutions may be used to distinguish between the precipitates due to chloride, bromide and iodide ions.

....

....

.... **(4 marks)**

*(c) Common table salt is sodium chloride. To prevent iodine deficiency, it may be 'iodised' by adding potassium iodide. Explain how you could adapt the methods described in part (b) to prepare a pure sample of silver iodide from a mixture of solid sodium chloride and sodium iodide.

....

....

....

.... **(6 marks)**

(Total for question 1 = 16 marks)

2 Compound **A** is a blue inorganic solid that contains one cation and one anion. It dissolves in deionised water to form a pink solution containing the complex ion **B**.

(a) Give the formulae of two cations that could be responsible for the blue colour of the solid.

.... **(2 marks)**

(b) The pink solution was divided into three portions.

(i) Concentrated ammonia solution was added dropwise to the first portion. A blue precipitate formed containing the species **C**, which re-dissolved in excess ammonia solution to form a pale brown solution containing a species **D**. When this solution was left to stand in air, its colour darkened due to the formation of species **E**.
Give the formulae of species **C**, **D** and **E**, and state why **D** darkens.

....

....

.... **(4 marks)**

(ii) Concentrated hydrochloric acid was added dropwise to the second portion. A blue solution containing the species **F** formed.
Give the formulae of species **F** and write an equation for the reaction that produces it from species **B**.

.... **(2 marks)**

(iii) A few drops of dilute hydrochloric acid were added to the third portion, followed by a few drops of barium chloride solution. A white precipitate of compound **G** formed.
Write an ionic equation, including state symbols, for the reaction that produces compound **G**.

....

.... **(2 marks)**

(c) Draw a diagram to show the shape of the complex ion **B**. State the bond angles between the ligands and the central metal ion, and the name the shape of the complex ion.

.... **(4 marks)**

(d) Give the formula of compound **A**.

.... **(1 mark)**

(Total for question 2 = 15 marks)

3 This question is about the kinetics of the hydrolysis reaction involving 2-bromo-2-methylpropane and aqueous hydroxide ions:

$(CH_3)_3CBr + OH^- \rightarrow (CH_3)_3COH + Br^-$

The rate of reaction was investigated using sodium hydroxide solution and an excess of 2-bromo-2-methylpropane. The time taken for all the hydroxide ions to be used up was measured for different initial concentrations of both reactants. The table below shows the results obtained.

Experiment	$[(CH_3)_3CBr]$ /mol dm^{-3}	$[OH^-]$ /mol dm^{-3}	Time for OH$^-$ to be used up /s	Initial rate /mol dm^{-3} s^{-1}
1	2.4×10^{-2}	1.6×10^{-3}	27	5.92×10^{-5}
2	1.2×10^{-2}	1.6×10^{-3}	56	2.86×10^{-5}
3	1.2×10^{-2}	2.4×10^{-3}	84	2.86×10^{-5}

(a) What was the smallest percentage error for a single time in the three experiments?

☐ **A** 3.7%

☐ **B** 1.8%

☐ **C** 1.2%

☐ **D** 0.6% **(1 mark)**

(b) Give the systematic (IUPAC) name for the organic product of the reaction.

.... **(1 mark)**

(c) Use the data in the table to deduce the orders of reaction with respect to $(CH_3)_3CBr$ and OH$^-$ ions. Justify your answers.

....

....

....

.... **(4 marks)**

(d) State the rate equation for the reaction.

.... **(1 mark)**

(e) Use your answer to part (d) and the results from Experiment 3 to calculate the rate constant. Give your answer to two significant figures, and state the units.

....

....

.... **(2 marks)**

(f) The mechanism for the reaction proceeds in two steps:
Step 1: $(CH_3)_3CBr \rightarrow (CH_3)_3C^+ + Br^-$
Step 2: $(CH_3)_3C^+ + OH^- \rightarrow (CH_3)_3COH$

(i) Explain which step is the rate-determining step.

....

.... **(2 marks)**

(ii) Draw the mechanism for the overall reaction.

.... **(2 marks)**

(iii) State whether it is an S_N1 mechanism or an S_N2 mechanism, and justify your answer.

....

.... **(1 mark)**

(Total for question 3 = 14 marks)

4 Hydrogen and iodine react together to form hydrogen iodide:

$$H_2(g) + I_2(g) \rightarrow 2HI(g)$$

The table below shows the results of an experiment in which the temperature was varied and the rate constant, k, determined at each temperature.

Temperature /K	$\frac{1}{T}$ /K^{-1}	ln k
689	1.45×10^{-3}	-3.95
694	1.44×10^{-3}	-3.74
700	1.43×10^{-3}	-3.52
709	1.41×10^{-3}	-3.10
714	1.40×10^{-3}	-2.89

The activation energy, E_a, of the reaction can be found using the equation:

$$\ln k = -\frac{E_a}{RT} + c$$

(a) Use the data in the table above to plot a graph of ln k against $\frac{1}{T}$. Draw a straight line of best fit through the points.

(2 marks)

(b) Calculate the gradient of the line of best fit.

....

.... **(2 marks)**

5

(c) Use your answer to part (b) to calculate the activation energy, E_a, for this reaction in kJ mol^{-1}. Give your answer to three significant figures. (R = 8.31 J K^{-1} mol^{-1})

....

.... **(2 marks)**

(Total for question 4 = 6 marks)

5 One of the stages in the manufacture of sulfuric acid by the Contact process involves the oxidation of sulfur dioxide in the presence of a vanadium(V) oxide catalyst:

$$SO_2(g) + \tfrac{1}{2}O_2(g) \rightleftharpoons SO_3(g)$$

The table shows some information about the substances involved.

Substance	$\Delta_f H^\circ$ /kJ mol^{-1}	S° /J K^{-1} mol^{-1}	Melting temperature /K	Boiling temperature /K
SO$_2$(g)	-297	248	200	263
O$_2$(g)		205	55	90
SO$_3$(l)	-441	96	290	318

(a) State the value for $\Delta_f H^\circ$ of O$_2$(g) and justify your answer.

....

.... **(2 marks)**

(b) (i) Calculate the standard enthalpy change of reaction, $\Delta_r H^\circ$, for the reaction between sulfur dioxide and oxygen using the data in the table and your answer to part (a).

....

.... **(1 mark)**

(ii) A data book gave this information for the reaction between sulfur dioxide and oxygen:

REACTION: $SO_2(g) + \tfrac{1}{2}O_2(g) \rightleftharpoons SO_3(g)$ $\Delta H^\circ = -98.5$ kJ mol^{-1}

Other than error, explain why the value calculated in part (b)(i) differs from this data book value.

....

....

.... **(2 marks)**

(c) Calculate the standard entropy change, ΔS°, for this reaction.

....

....

.... **(1 mark)**

(d) Use the data book value for ΔH° and your value for ΔS° calculated in part (c) to explain, by calculation, whether the reaction is feasible at 298 K.

....

....

.... **(3 marks)**

6

(e) Explain, in terms of entropy, why this reaction is not feasible at very high temperatures.

....

....

.... **(2 marks)**

(f) Explain the effect on the position of equilibrium in this reaction of increasing:

(i) the temperature

....

.... **(2 marks)**

(ii) the pressure

....

....

.... **(2 marks)**

(Total for question 5 = 15 marks)

6 The bark of the English willow tree, *Salix alba*, contains a natural analgesic (painkiller). This is salicylic acid, 2-hydroxybenzoic acid. Acetylsalicylic acid or 2-ethanoyloxybenzoic acid, better known as aspirin, is an analgesic with fewer side effects.

2-hydroxybenzoic acid reacts with ethanoic anhydride in the presence of a phosphoric acid catalyst to form 2-ethanoyloxybenzoic acid and ethanoic acid:

(a) Calculate the atom economy of this process.

....

.... **(2 marks)**

(b) Describe how you could purify aspirin by recrystallisation using water.

....

....

....

.... **(4 marks)**

(c) Aspirin passes unchanged through the acidic conditions in the stomach, but it is hydrolysed in the alkaline conditions of the small intestine.

$$CH_3COOC_6H_4COOH + 2OH^- \rightarrow HOC_6H_4COO^- + CH_3COO^- + H_2O$$

A student carried out an experiment, using alkaline hydrolysis, to determine the percentage purity of aspirin in aspirin tablets. This is the method used.

7

---

1. Crushed and weighed aspirin tablets were hydrolysed by heating for 15 minutes with 50.0 cm^3 (an excess) of 1.00 mol dm^{-3} sodium hydroxide solution.
2. The resulting solution was transferred to a 250 cm^3 volumetric flask, with washings, and made up to the mark with deionised water.
3. 25 cm^3 portions of the solution from step 2 were titrated with 0.150 hydrochloric acid using phenolphthalein indicator.

The results are shown in the table below.

Mass of crushed aspirin tablets /g	Mean titre /cm^3
1.52	23.60

(i) Calculate the amount, in mol, of sodium hydroxide added to the crushed aspirin tablets in Step 1.

....

.... **(1 mark)**

(ii) Use the mean titre to calculate the amount, in mol, of sodium hydroxide that remained after the hydrolysis of the crushed aspirin tablets.

....

....

.... **(3 marks)**

(iii) Calculate the amount of sodium hydroxide used in the hydrolysis.

....

.... **(1 mark)**

(iv) Use your answer to part (iii) to calculate the amount of aspirin that was hydrolysed in the experiment.

....

.... **(1 mark)**

(v) Calculate the percentage purity of aspirin in the tablets. Give your answer to three significant figures.

....

....

.... **(2 marks)**

(Total for question 6 = 14 marks)

7 The table below shows the standard electrode (redox) potentials, E°, for some half-cell reactions.

Half-cell reaction	E° /V
$H^+ + e^- \rightleftharpoons \tfrac{1}{2}H_2$	0.00
$\tfrac{1}{2}O_2 + H_2O + 2e^- \rightleftharpoons 2OH^-$	$+0.40$
$\tfrac{1}{2}O_2 + 2H^+ + 2e^- \rightleftharpoons H_2O$	$+1.23$
$\tfrac{1}{2}Cr_2O_7^{2-} + 7H^+ + e^- \rightleftharpoons Cr^{3+} + 3\tfrac{1}{2}H_2O$	$+1.33$
$\tfrac{1}{2}Cl_2 + e^- \rightleftharpoons Cl^-$	$+1.36$
$MnO_4^- + 8H^+ + 5e^- \rightleftharpoons Mn^{2+} + 4H_2O$	$+1.51$

8

(a) Which species in the table is the most powerful reducing agent?

☐ **A** H^+

☐ **B** H_2

☐ **C** MnO_4^-

☐ **D** Mn^{2+} **(1 mark)**

(b) The standard hydrogen electrode is the primary reference electrode.

(i) Explain why a reference electrode is necessary for measuring electrode potentials.

....

....
    (2 marks)

(ii) Sketch a labelled diagram to show the essential features of the standard hydrogen electrode, and include the conditions used.

    (6 marks)

(iii) Explain the purpose of a salt bridge when measuring electrode potentials.

....

....
    (2 marks)

(c) Potassium dichromate(VI), $K_2Cr_2O_7$(aq), and potassium manganate(VII), $KMnO_4$(aq), may be used in redox titrations, but they must be acidified using a dilute acid. Explain, with reference to standard electrode potentials, why potassium dichromate(VI) solution may be acidified with dilute hydrochloric acid, but this acid is not suitable for acidifying potassium manganate(VII) solution.

....

....

....

....

....
    (3 marks)

(d) Acidic or alkaline electrolytes may be used in hydrogen–oxygen fuel cells.

(i) Write half-equations for the reactions that occur at each electrode in a hydrogen–oxygen fuel cell operating under acidic conditions.

....

....
    (2 marks)

(ii) Calculate the standard emf, $E^{\ominus}_{cell}$, for a hydrogen–oxygen fuel cell operating under acidic conditions.

....

....
    (1 mark)

9

(iii) The overall reaction for a hydrogen–oxygen fuel cell operating under alkaline conditions is the same as for one operating under acidic conditions, and hydrogen gas is still supplied to the negative electrode.

Deduce the standard electrode potential, $E^{\ominus}$, for the negative electrode in a hydrogen–oxygen fuel cell operating under alkaline conditions. Write the half-equation for the reaction that occurs.

....

....
    (3 marks)

    (Total for question 7 = 20 marks)

8   This question is about different methods of analysis.
The table below gives information about four substances with $M_r = 60.0$

Substance	Structural formula
propan-1-ol	$CH_3CH_2CH_2OH$
propan-2-ol	$CH_3CH(OH)CH_3$
ethanoic acid	CH_3COOH
ethyl methyl ether	$CH_3CH_2OCH_3$

(a) State two advantages of instrumental methods of analysis, such as mass spectrometry, over simple test tube methods of analysis.

....

....
    (2 marks)

(b) The mass spectrometer can provide accurate M_r values to four decimal places. Explain why this allows you to distinguish between ethanoic acid and the other three compounds, but not between those three compounds.

....

....
    (2 marks)

(c) The peak with the highest relative intensity in the mass spectrum of propan-1-ol is at $m/z = 31$, but in the mass spectrum of propan-2-ol it is at $m/z = 45$. Give the structural formulae of the fragments most likely to be responsible for these peaks.

....

....
    (2 marks)

(d) State which two compounds would produce ^{13}C NMR spectra with just two peaks.

....
    (1 mark)

(e) Predict the high-resolution splitting pattern of the ^1H atoms in ethyl methyl ether, giving reasons for your answer.

....

....

....
    (3 marks)

10

(f) The table below gives information from the high–resolution ^1H NMR spectrum of one of these four compounds.

Chemical shift, δ /ppm	Integration ratio	Splitting pattern
4.00	1	septet
2.15	1	singlet
1.21	6	doublet

Deduce which of the four compounds produced this spectrum, and explain your answer.

....

....

....
    (4 marks)

*(g) The simplified infrared spectra of two of these compounds are shown below.

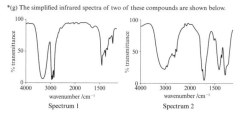

Spectrum 1   Spectrum 2

The table below shows some infrared absorption wavenumbers.

Group	Wavenumber range /cm^{-1}
O–H (in alcohols)	3750–3200
O–H (in carboxylic acids)	3300–2500
C–H (stretching)	2962–2853
C=O	1725–1700
C–H (bending)	1485–1365

Explain fully using information from the spectra and table which compound(s) could be responsible for each spectrum.

....

....

....

....

....
    (6 marks)

    (Total for question 8 = 20 marks)

         TOTAL FOR PAPER =120 MARKS

11

Periodic Table

Group

Key

Atomic (proton number)
Atomic symbol
Name
Relative atomic mass

							1 **H** Hydrogen 1.0										

Period

1 (1)	2 (2)	(3)	(4)	(5)	(6)	(7)	(8)	(9)	(10)	(11)	(12)	3 (13)	4 (14)	5 (15)	6 (16)	7 (17)	8 (18)
																	2 **He** Helium 4.0
3 **Li** Lithium 6.9	4 **Be** Beryllium 9.0											5 **B** Boron 10.8	6 **C** Carbon 12.0	7 **N** Nitrogen 14.0	8 **O** Oxygen 16.0	9 **F** Fluorine 19.0	10 **Ne** Neon 20.2
11 **Na** Sodium 23.0	12 **Mg** Magnesium 24.3											13 **Al** Aluminium 27.0	14 **Si** Silicon 28.1	15 **P** Phosphorus 31.0	16 **S** Sulfur 32.1	17 **Cl** Chlorine 35.5	18 **Ar** Argon 39.9
19 **K** Potassium 39.1	20 **Ca** Calcium 40.1	21 **Sc** Scandium 45.0	22 **Ti** Titanium 47.9	23 **V** Vanadium 50.9	24 **Cr** Chromium 52.0	25 **Mn** Manganese 54.9	26 **Fe** Iron 55.8	27 **Co** Cobalt 58.9	28 **Ni** Nickel 58.7	29 **Cu** Copper 63.5	30 **Zn** Zinc 65.4	31 **Ga** Gallium 69.7	32 **Ge** Germanium 72.6	33 **As** Arsenic 74.9	34 **Se** Selenium 79.0	35 **Br** Bromine 79.9	36 **Kr** Krypton 83.8
37 **Rb** Rubidium 85.5	38 **Sr** Strontium 87.6	39 **Y** Yttrium 88.9	40 **Zr** Zirconium 91.2	41 **Nb** Niobium 92.9	42 **Mo** Molybdenum 95.9	43 **Tc** Technetium (98)	44 **Ru** Ruthenium 101.1	45 **Rh** Rhodium 102.9	46 **Pd** Palladium 106.4	47 **Ag** Silver 107.9	48 **Cd** Cadmium 112.4	49 **In** Indium 114.8	50 **Sn** Tin 118.7	51 **Sb** Antimony 121.8	52 **Te** Tellurium 127.6	53 **I** Iodine 126.9	54 **Xe** Xenon 131.3
55 **Cs** Caesium 132.9	56 **Ba** Barium 137.3	57 **La*** Lanthanum 138.9	72 **Hf** Hafnium 178.5	73 **Ta** Tantalum 180.9	74 **W** Tungsten 183.8	75 **Re** Rhenium 186.2	76 **Os** Osmium 190.2	77 **Ir** Iridium 192.2	78 **Pt** Platinum 195.1	79 **Au** Gold 197.0	80 **Hg** Mercury 200.6	81 **Tl** Thallium 204.4	82 **Pb** Lead 207.2	83 **Bi** Bismuth 209.0	84 **Po** Polonium (209)	85 **At** Astatine (210)	86 **Rn** Radon (222)
87 **Fr** Francium (223)	88 **Ra** Radium (226)	89 **Ac*** Actinium (227)	104 **Rf** Rutherfordium (261)	105 **Db** Dubnium (262)	106 **Sg** Seaborgium (266)	107 **Bh** Bohrium (264)	108 **Hs** Hassium (277)	109 **Mt** Meitnerium (268)	110 **Ds** Darmstadtium (271)	111 **Rg** Roengenium (272)	112 **Cn** Copernicium 112	Fl flerovium 114		Lv livermorium 116		

58 **Ce** Cerium 140.1	59 **Pr** Praseodymium 140.9	60 **Nd** Neodymium 144.2	61 **Pm** Promethium 144.9	62 **Sm** Samarium 150.4	63 **Eu** Europium 152.0	64 **Gd** Gadolinium 157.2	65 **Tb** Terbium 158.9	66 **Dy** Dysprosium 162.5	67 **Ho** Holmium 164.9	68 **Er** Erbium 167.3	69 **Tm** Thulium 168.9	70 **Yb** Ytterbium 173.0	71 **Lu** Lutetium 175.0
90 **Th** Thorium 232.0	91 **Pa** Protactinium (231)	92 **U** Uranium 238.1	93 **Np** Neptunium (237)	94 **Pu** Plutonium (242)	95 **Am** Americium (243)	96 **Cm** Curium (247)	97 **Bk** Berkelium (245)	98 **Cf** Californium (251)	99 **Es** Einsteinium (254)	100 **Fm** Fermium (253)	101 **Md** Mendelevium (256)	102 **No** Nobelium (254)	103 **Lr** Lawrencium (257)

Data booklet

Physical constants

Avogadro constant (L)	$6.02 \times 10^{23}\ \text{mol}^{-1}$
Elementary charge (e)	$1.60 \times 10^{-19}\ \text{C}$
Gas constant (R)	$8.31\ \text{J mol}^{-1}\text{K}^{-1}$
Molar volume of ideal gas:	
at r.t.p.	$24\ \text{dm}^3\text{mol}^{-1}$
Specific heat capacity of water	$4.18\ \text{J g}^{-1}\text{K}^{-1}$
Ionic product of water (K_w)	$1.00 \times 10^{-14}\ \text{mol}^2\text{dm}^{-6}$

$1\ \text{dm}^3 = 1\ 000\ \text{cm}^3 = 0.001\ \text{m}^3$

Correlation of infrared absorption wavenumbers with molecular structure

Group	Wavenumber range/cm^{-1}
C—H stretching vibrations	
Alkane	2962–2853
Alkene	3095–3010
Alkyne	3300
Arene	3030
Aldehyde	2900–2820 and 2775–2700
C—H bending variations	
Alkane	1485–1365
Arene 5 adjacent hydrogen atoms	750 and 700
4 adjacent hydrogen atoms	750
3 adjacent hydrogen atoms	780
2 adjacent hydrogen atoms	830
1 adjacent hydrogen atom	880
N—H stretching vibrations	
Amine	3500–3300
Amide	3500–3140
O—H stretching vibrations	
Alcohols and phenols	3750–3200
Carboxylic acids	3300–2500
C=C stretching vibrations	
Isolated alkene	1669–1645
Arene	1600, 1580, 1500, 1450
C=O stretching vibrations	
Aldehydes, saturated alkyl	1740–1720
Ketones alkyl	1720–1700
Ketones aryl	1700–1680
Carboxylic acids alkyl	1725–1700
aryl	1700–1680
Carboxylic acid anhydrides	1850–1800 and 1790–1740
Acyl halides chlorides	1795
bromides	1810
Esters, saturated	1750–1735
Amides	1700–1630
Triple bond stretching vibrations	
CN	2260–2215
CC	2260–2100

¹H nuclear magnetic resonance chemical shifts relative to tetramethylsilane (TMS)

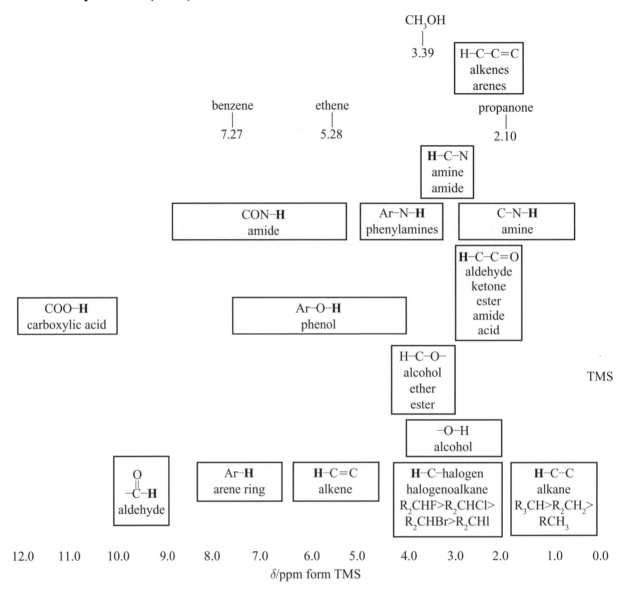

¹³C nuclear magnetic resonance chemical shifts relative to tetramethylsilane (TMS)

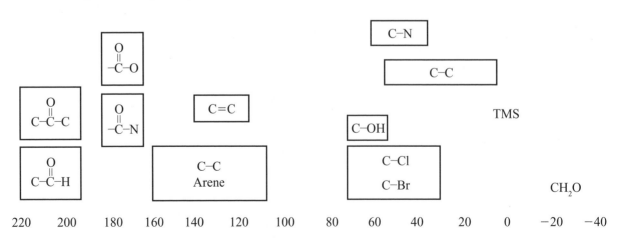

Pauling electronegativity index

							H										He
							2·1										
Li	Be											B	C	N	O	F	Ne
1·0	1·5											2·0	2·5	3·0	3·5	4·0	
Na	Mg											Al	Si	P	S	Cl	Ar
0·9	1·2											1·5	1·9	2·1	2·5	3·0	
K	Ca	Sc	Ti	V	Cr	Mn	Fe	Co	Ni	Cu	Zn	Ga	Ge	As	Se	Br	Kr
0·8	1·0	1·3	1·5	1·6	1·6	1·5	1·8	1·8	1·8	1·9	1·6	1·6	2·0	2·0	2·4	2·8	
Rb	Sr	Y	Zr	Nb	Mo	Tc	Ru	Rh	Pd	Ag	Cd	In	Sn	Sb	Te	I	Xe
0·8	1·0	1·2	1·3	1·6	2·1	1·9	2·2	2·2	2·2	1·9	1·6	1·7	1·9	1·9	2·1	2·5	
Cs	Ba	La	Hf	Ta	W	Re	Os	Ir	Pt	Au	Hg	Tl	Pb	Bi	Po	At	Rn
0·7	0·9	1·1	1·3	1·5	2·3	1·9	2·2	2·2	2·2	2·5	2·0	1·6	1·8	1·9	2·0	2·2	

Relation in electronegativity difference, ΔNe and ionic character P/%

Electronegativity difference ΔN_e	0·1	0·3	0·5	0·7	1·0	1·3	1·5	1·7	2·0	2·5	3·0
Percentage ionic character $P/\%$	0·5	2	6	12	22	34	43	51	63	79	89

Indicators

		pKin (at 298 K)	*acid*	pH range	*alkaline*
1	Thymol blue (acid)	1.7	red	1.2−2.8	yellow
2	Screened methyl orange	3.7	purple	3.2−4.2	green
3	Methyl orange	3.7	red	3.2−4.4	yellow
4	Bromophenol blue	4.0	yellow	2.8−4.6	blue
5	Bromocresol green	4.7	yellow	3.8−5.4	blue
6	Methyl red	5.1	red	4.2−6.3	yellow
7	Litmus		red	5.0−8.0	blue
8	Bromothymol blue	7.0	yellow	6.0−7.6	blue
9	Phenol red	7.9	yellow	6.8−8.4	red
10	Phenolphthalein (in ethanol)	9.3	colourless	8.2−10.0	red

Standard electrode potentials

$E^{\ominus}$ Standard electrode potential of aqueous system at 298 K, that is, standard emf of electrochemical cell in the hydrogen half-cell forms the left-hand side electrode system.

	Right-hand electrode system	$E^{\ominus}/V$
1	$Na^+ + e^- \rightleftharpoons Na$	-2.71
2	$Mg^{2+} + 2e^- \rightleftharpoons Mg$	-2.37
3	$Al^{3+} + 3e^- \rightleftharpoons Al$	-1.66
4	$V^{2+} + 2e^- \rightleftharpoons V$	-1.18
5	$Zn^{2+} + 2e^- \rightleftharpoons Zn$	-0.76
6	$Cr^{3+} + 3e^- \rightleftharpoons Cr$	-0.74
7	$Fe^{2+} + 2e^- \rightleftharpoons Fe$	-0.44
8	$Cr^{3+} + e^- \rightleftharpoons Cr^{2+}$	-0.41
9	$V^{3+} + e^- \rightleftharpoons V^{2+}$	-0.26
10	$Ni^{2+} + 2e^- \rightleftharpoons Ni$	-0.25
11	$H+ + e^- \rightleftharpoons \frac{1}{2}H_2$	0.00
12	$\frac{1}{2}S_4O_6^{2-} + e^- \rightleftharpoons S_2O_3^{2-}$	$+0.09$
13	$Cu^{2+} + e^- \rightleftharpoons Cu^+$	$+0.15$
14	$Cu^{2+} + 2e^- \rightleftharpoons Cu$	$+0.34$
15	$VO^{2+} + 2H^+ + e^- \rightleftharpoons V^{3+} + H_2O$	$+0.34$
16	$\frac{1}{2}O_2 + H_2O + 2e^- \rightleftharpoons 2OH^-$	$+0.40$
17	$S_2O_3^{2-} + 6H^+ + 4e^- \rightleftharpoons 2S + 3H_2O$	$+0.47$
18	$Cu^+ + e^- \rightleftharpoons Cu$	$+0.52$
19	$\frac{1}{2}I_2 + e^- \rightleftharpoons I^-$	$+0.54$
20	$3O_2 + 2H^+ + 2e^- \rightleftharpoons H_2O_2$	$+0.68$
21	$Fe^{3+} + e^- \rightleftharpoons Fe^{2+}$	$+0.77$
22	$Ag^+ + e^- \rightleftharpoons Ag$	$+0.80$
23	$NO_3^- + 2H^+ + e^- \rightleftharpoons NO_2 + H_2O$	$+0.80$
24	$ClO^- + H_2O + 2e^- \rightleftharpoons Cl^- + 2OH^-$	$+0.89$
25	$VO2^+ + 2H^+ + e^- \rightleftharpoons VO^{2+} + H_2O$	$+1.00$
26	$\frac{1}{2}Br_2 + e^- \rightleftharpoons Br^-$	$+1.09$
27	$\frac{1}{2}O_2 + 2H^+ + 7H^+ + 3e^- \rightleftharpoons Cr^{3+} + \frac{7}{2}H_2O$	$+1.23$
28	$\frac{1}{2}Cr_2O_7^{2-} + 2e^- \rightleftharpoons H_2O$	$+1.33$
29	$\frac{1}{2}Cl_2 + e^- \rightleftharpoons Cl^-$	$+1.36$
30	$MnO4^- + 8H^+ + 5e^- \rightleftharpoons Mn^{2+} + 4H_2O$	$+1.51$
31	$\frac{1}{2}H_2O_2 + H_+ + e^- \rightleftharpoons H_2O$	$+1.77$

ANSWERS

You will find some advice next to some of the answers. This is written in *italics*. It is not part of the mark scheme but gives you a little more information.

1. Atomic structure and isotopes

1 proton: relative mass = 1
   relative charge = +1 **(1)**
   neutron: relative mass = 1
   relative charge = 0 **(1)**
   electron: relative mass = $\frac{1}{1840}$
   relative charge = −1 **(1)** *Allow $\frac{1}{1800} - \frac{1}{2000}$*

2 B **(1)**

3 (a) (weighted) mean mass of an atom of an element **(1)**
     compared to $\frac{1}{12}$th the mass of a ^{12}C atom **(1)**
   (b) (Isotopes are atoms that have the same number of)
     protons (and electrons) **(1)**
     but different numbers of neutrons **(1)**.

4 atomic number = 16 **(1)**
   mass number = 33 **(1)**
   element = sulfur, S **(1)**

5 A **(1)** *Both have 16 electrons*

2. Mass spectrometry

1 $(^{14}N-^{15}N)^+$ $m/z = 29$ **(1)**
   $(1^{15}N-^{15}N)^+$ $m/z = 30$ **(1)**

2 D **(1)**

3 $[(75.78 \times 35) + (24.22 \times 37)] \div 100$ **(1)**
   $= 35.48$ **(1)**

4 $\frac{(58 \times 68.1) + (60 \times 26.2) + (61 \times 1.1) + (62 \times 3.7) + (64 \times 0.9)}{100}$ **(1)**
   $= 58.8$ **(1)**   (58.759 before rounding)

5 The peak with the largest m/z value (ignoring a small M+1 peak if present) is the molecular ion peak. **(1)**
   Its m/z value is equal to the relative molecular mass of the molecule. **(1)**

3. Shells, sub-shells and orbitals

1 (An orbital is a region) within an atom that can hold up to two electrons with opposite spins. **(1)**

2 (a) spin **(1)**
   (b)

     s-orbital   p-orbital
   1 mark for each correct diagram
   Axes need not be shown in the p-orbital diagram

3 *All three correct for 1 mark:*

Sub-shell	s	p	d
Maximum number of electrons	2	6	10

4 C **(1)**

5 B **(1)**

4. Electronic configurations

1 (a) $1s^2\ 2s^2\ 2p^6\ 3s^1$ **(1)**
   (b) s-block **(1)**

2 B **(1)**

3 C **(1)**

4 (a) $1s^2\ 2s^2\ 2p^6\ 3s^2\ 3p^6\ 3d^{10}\ 4s^1$ **(1)**
     $1s^2\ 2s^2\ 2p^6\ 3s^2\ 3p^6\ 4s^1\ 3d^{10}$ is also accepted
   (b) d-block **(1)**

5 Magnesium is $1s^2\ 2s^2\ 2p^6\ 3s^2$ and calcium is $1s^2\ 2s^2\ 2p^6\ 3s^2\ 3p^6\ 4s^2$ / both have two s electrons in their highest occupied orbital. **(1)**
   The electronic configuration of an element determines its chemical properties. **(1)**

6 D **(1)**

5. Ionisation energies

1 (a) energy change / enthalpy change per mole **(1)**
     to remove an electron **(1)**
     from gaseous atoms **(1)**
   (b) $Al^{2+}(g) \rightarrow Al^{3+}(g) + e^-$ **(1)**
     State symbols are needed to gain the mark.

2 B **(1)**

3 (a) (There is a large jump between the) 6th and 7th electrons removed. **(1)**
   (b) Answer should include any two of these points for
     1 mark each:
     The charge on the ion increases.
     Fewer electrons are attracted by the same number of protons.
     Electrons removed are closer to the nucleus.

4 Shielding by inner completed shells increases. **(1)**
   The outer electron is further from the nucleus / more occupied shells / radius increases. **(1)**
   Force of attraction between nucleus and outer electron decreases. **(1)**

5 The number of protons increases / nuclear charge increases. **(1)**
   Negligible difference in shielding (because electrons enter the same shell). **(1)**
   Outer electron is closer to the nucleus / radius decreases. **(1)**
   Force of attraction between nucleus and outer electron increases. **(1)**

6. Periodicity

1 (a) trend in the properties of elements going across a period **(1)**
     Trends repeated in the next period / elements in the same group have similar properties. **(1)**
   (b) decreases **(1)**

2 (a) (It has a giant) lattice / atomic **(1)**
     (structure held together by) covalent / very strong **(1)**
     (bonds. A lot of energy is needed to) break these bonds / separate the atoms. **(1)**
   (b) Aluminium has a giant lattice / metallic structure **(1)**
     with metallic bonding. **(1)**
     A lot of energy is needed to break these bonds. **(1)**
   (c) The bonds must be very weak (because the boiling point is so low). **(1)**

3 (a) cross higher than Na and Al, but lower than Si **(1)**

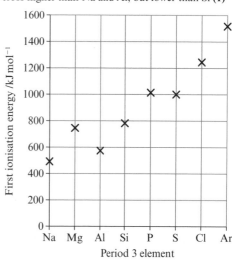

(b) Higher because:
  oxygen has fewer electron shells / smaller atomic radius / smaller atoms than sulfur. **(1)**
  Outer electron is closer to the nucleus in oxygen / less shielded / removed from a lower energy level than in sulfur. **(1)**
  Opposite argument for sulfur is acceptable.

7. Exam skills 1

1 (a) (i) $1s^2 2s^2 2p^6 3s^2 3p^3$ **(1)**
    (ii) $1s^2 2s^2 2p^6 3s^2 3p^6 3d^5$ **(1)**
  (b) $\frac{(54 \times 5.85) + (56 \times 91.75) + (57 \times 2.12) + (58 \times 0.28)}{100}$ **(1)**
    $= 55.91$ **(1)**
  (c) $Fe^+(g) \rightarrow Fe^{2+}(g) + e^-$ / $Fe^+(g) + e^- \rightarrow Fe^{2+}(g) + 2e^-$ **(1)**
  (d) (i) First ionisation energy generally increases
      greater nuclear charge / more protons / proton number increases **(1)**
      outer electrons are in the same shell / same energy level / have similar shielding / atomic radius decreases / outer electron closer to the nucleus **(1)**
      greater force of attraction between the nucleus and outer electrons / outer electrons more strongly held by the nucleus **(1)**
    (ii) First ionisation energy of sulfur is less than that of phosphorus, the element before it. **(1)**
      Electron removed from a pair of electrons in the 3p sub-shell / pairing has happened in the 3p sub-shell / suitable diagram to explain this. **(1)**
      3p $\uparrow\downarrow$ $\uparrow$ $\uparrow$
      There is repulsion between the paired electrons. **(1)**

8. Ions

1 **A (1)**
2 **C (1)**
3 (a) Correct number of circles and electrons shown, with correct charge. **(1)**

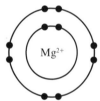

  (b) Correct number of circles and electrons shown, with correct charge. **(1)**
    Dots or crosses may be used. Circles do not need to be present but the electrons must clearly be in their correct shells. Note that these ions are isoelectronic.

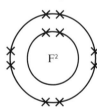

4 positive electrode: purple colour (due to manganate(VII) ions)
  negative electrode: no colour / no change (potassium ions are colourless)
  *Both must be correct for **1** mark.*
5 (a) Each copper atom loses two electrons. **(1)**
    $Cu \rightarrow Cu^{2+} + 2e^-$ **(1)**
  (b) Atoms gain one or more electrons. **(1)**

9. Ionic bonds

1 (a) (It is the) strong electrostatic force **(1)** (of attraction between) oppositely charged ions. **(1)**
  (b) Force between ions with like charges / between nuclei / between electrons **(1)**
2 **C (1)**
3 *Answer should include:*
  • The aluminium ion is smaller / has a smaller ionic radius. **(1)**
  • The aluminium ion has a higher charge. **(1)**
  The reverse argument for the sodium ion is also acceptable.
4 *Answer should include:*
  • Lithium ions are smaller than potassium ions. **(1)**
  • Lithium ions and potassium ions have the same charge / lithium ions have a greater charge density. **(1)**

5 + and − signs in the correct places **(1)**

6 **A (1)**

10. Covalent bonds

1 The strong electrostatic force of attraction **(1)** between two nuclei and the shared pair of electrons between them **(1)**.
2 (a) covalent bond **(1)**
  (b) (i) dative covalent bond / coordinate bond **(1)**
    (ii) Lone pair of electrons on H / H⁻ ion **(1)** donated to vacant orbital on Al atom. **(1)** (Both electrons come from H / H⁻, for **1** mark only.)
3 (a)

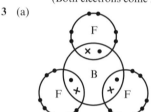

    1 mark for 3 correct bonding pairs.
    1 mark for correct lone pairs on F atoms only.
  (b)

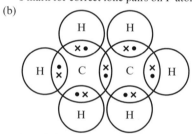

    1 mark for C−C bonding pair.
    1 mark for correct C−H bonding pairs.
4 **B (1)**

11. Covalent bond strength

1 **A (1)**
2 **D (1)**
3 **D (1)**
4 (a)

    1 mark for correct bonding pairs.
    1 mark for correct lone pairs.
  (b)

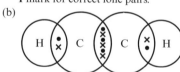

    1 mark for 3 correct bonding pairs between carbon atoms.
    1 mark for correct bonding pairs between carbon atoms and hydrogen atoms.
5

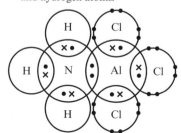

    1 mark for dative pair between N and Al.
    1 mark for correct bonding pairs N−H and Al−Cl.
    1 mark for correct lone pairs on Cl atoms only.

12. Shapes of molecules and ions

1 (a)   180°   correct shape **(1)** correct angle **(1)**

$$O=C=O$$

(b)   H   correct shape **(1)** correct angle **(1)**

109.5° C····H

H   H

(c)   Cl   correct shape **(1)** correct angle **(1)**

120°   Al

Cl   Cl

(d)   F   correct shape **(1)** both correct angles **(1)**

90°   F

F—P   120°

F

2 A **(1)**

3 (a) tetrahedral **(1)** 109.5° **(1)**
   There are 4 bonding pairs which repel equally. **(1)**
   (b) trigonal pyramidal **(1)** 107° **(1)**
   There are 3 bonding pairs and one lone pair. **(1)**
   (c) bent line / V shape **(1)** 104.5° **(1)**
   There are 2 bonding pairs and two lone pairs. **(1)**

4 D **(1)**

13. Electronegativity and bond polarity

1 B **(1)**

2 A **(1)**

3 (a) (i) δ^+ means **partial** positive charge / δ^- means **partial** negative charge. **(1)**
   (ii) δ^- **(1)** because oxygen is more electronegative (than carbon). **(1)**
   (b) It has two identical dipoles. **(1)**
   In a straight line / dipoles cancel out / centre of positive charge coincides with centre of negative charge. **(1)**

4 stream of ethanol (e.g. from a burette) **(1)**
   bring a charged object near (e.g. balloon, rod) **(1)**
   stream will be deflected **(1)**
   An alternative answer involves adding to a polar solvent (1), e.g. water (1), ethanol completely dissolves / no layers formed (1).

5 D **(1)**

14. London forces

1 C **(1)**

2 (a) London force **(1)**
   Instantaneous dipole / temporary dipole in one molecule **(1)** produces an induced dipole in another molecule. **(1)**
   (b) (Pentane has the greatest) number of points of contact between molecules **(1)** so its intermolecular forces are strongest (and take the most energy to break) **(1)**.

3 (a) increases **(1)**
   (b) As the chain length increases, the M_r increases **(1)**.
   The number of electrons increases. **(1)**
   The strength of the intermolecular forces / London forces increases. **(1)**

15. Permanent dipoles and hydrogen bonds

1 D **(1)**

2 (a) London forces **(1)**
   Increase in strength as the number of electrons increase. **(1)**
   (b) 170 K (± 2 K) **(1)**
   (c) Nitrogen is more electronegative (than P / As / Sb). **(1)**
   N–H bond is more polar (than P–H / As–H / Sb–H bond). **(1)**
   Hydrogen bonds form between lone pair on N and δ^+ on H. **(1)**

3 (a) δ^- O δ^+

H   H δ^- O δ^+

H   H

Dashed line / dotted line between H on left hand H_2O and lone pair on O on next H_2O. **(1)**

O–H---O in a straight line **(1)**
Partial charges need not be shown.
An alternative answer is to show O---H–O to water drawn at top right.

(b) There are hydrogen bonds between the molecules of both compounds **(1)**
but ammonia forms one bond on average and water forms two bonds on average **(1)**.
More energy is needed to separate water molecules than to separate ammonia molecules (so water has a higher boiling point). **(1)**

16. Choosing a solvent

1 B **(1)**

2   H_3C δ^- O δ^+

CH_2 H. δ^- O δ^+

H   H

Dashed line / dotted line between H on ethanol and lone pair on O of water **(1)**
O–H---O in a straight line. **(1)**
Partial charges need not be shown.
An alternative answer is to show O---H–O from ethanol to water drawn at top right.

3 (a) (i)   London forces **(1)**
   (ii) hydrogen bonds **(1)**
In (b) and (c) the answers should consider the relative strength of the attractive forces, such as those between:
   • *alcohol molecules*
   • *water molecules*
   • *alcohol molecules and water molecules.*
   (b) The London forces between the carbon–carbon chains in 2-methylpropan-2-ol are weaker (than those in the other alcohols). **(1)**
   The London forces between 2-methylpropan-2-ol and water are weaker (than those between the other alcohols and water). **(1)**
   The hydrogen bonds between 2-methylpropan-2-ol and water are stronger (than those between the other alcohols and water). **(1)**
   (c) butan-1-ol **(1)**
   There are only London forces between cyclohexane molecules / no hydrogen bonds. **(1)**
   Stronger (London) forces form between butan-1-ol and cyclohexane than between the other alcohols and cyclohexane. **(1)**

17. Giant lattices

1 D **(1)**

2 (a) (strong) electrostatic force of attraction between metal ions / positive ions / cations and delocalised electrons / sea of electrons **(1)**
   (b) giant lattice **(1)** of metal ions **(1)** with delocalised electrons **(1)**

3 (a) giant lattice **(1)**
   covalent bonds between carbon atoms **(1)**
   each carbon atom bonded to four other carbon atoms / tetrahedral arrangement **(1)**
   (b) giant lattice **(1)**
   each carbon atom covalently bonded to three other carbon atoms to form layers **(1)**
   London forces between layers **(1)**
   delocalised electrons **(1)**

4 Ions shown as + and −, or Na^+ and Cl^-, and placed correctly. **(1)**, e.g.

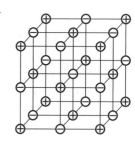

18. Structure and properties

1 **C (1)**

2 (a) It contains delocalised electrons. **(1)**

(b) Graphite contains delocalised electrons but diamond does not. **(1)**

(c) It contains ions / Na^+ and Cl^- ions. **(1)**
These are free to move around when molten or in solution, but not when solid. **(1)**

3 (a) Both have covalent bonds (between their atoms). **(1)**

(b) Iodine has a simple molecular structure and diamond has a giant lattice structure. **(1)**

(c) There are weak London forces between iodine molecules. **(1)**
There are many strong covalent bonds that must be broken in diamond. **(1)**

4 Simple molecular **(1)**
with covalent bonds (between atoms) and London forces (between molecules). **(1)**
The low temperature for sublimation shows that there are weak forces (between molecules). **(1)**

19. Exam skills 2

1 (a) Answer should include:
- electrostatic forces of attraction between metal ions / positive ions / cations **(1)** and
- delocalised electrons / sea of electrons **(1)**
- metallic bonds are strong / need a lot of energy to overcome **(1)**

(b) Answer should include **three** of the following points for **1** mark each:
- giant
- lattice
- covalent bonds
- (covalent bonds) are strong / need a lot of energy to overcome

(c) (i) diagram **(1)**   (ii) diagram **(1)**

Bond angle: 109.5° **(1)**   Bond angle: 90° **(1)**
Tetrahedral **(1)**   Square planar **(1)**

(d) (i) Hydrogen fluoride has hydrogen bonding between its molecules and hydrogen chloride does not. **(1)**
Hydrogen bonding is stronger than other intermolecular forces / hydrogen chloride only has London forces and weak permanent dipole−permanent dipole forces between its molecules **(1)**
more energy needed to separate the hydrogen fluoride molecules **(1)**.

(ii) London forces between hydrogen iodide molecules greater than those between hydrogen bromide molecules **(1)**
because hydrogen iodide molecules have more electrons / are larger molecules (than hydrogen chloride) **(1)**
more energy needed to separate the hydrogen iodide molecules / permanent dipole in hydrogen iodide is weaker than that in hydrogen bromide / increase in strength of London forces (in hydrogen iodide) greater than decrease in the strength of the permanent dipole **(1)**

20. Oxidation numbers

1 (a) 0 **(1)**   (b) −2 **(1)**   (c) +3 **(1)**

2 **C (1)**

3 **C (1)**

4 (a) Cl_2O <u>and</u> ClO^- **(1)**   (b) Chlorine(VI) oxide **(1)** +1 **(1)**

5 (a) Cu_2S **(1)**   (b) Cu_2Cl_2 **(1)**   (c) VF_3 **(1)**
(d) NO_2 **(1)**   (e) N_2O_5 **(1)**   (f) CrO_3 **(1)**

21. Redox reactions

1 **A (1)**

2 **A (1)**

3 (a) a species / substance that gains electrons **(1)**

(b) reduction **(1)** because the oxidation number decreases from +6 to +3 **(1)**

4 Aluminium is oxidised **(1)** and iron oxide / iron is reduced. **(1)**

5 (a) $HgCl_2$ +2 **(1)**
Hg 0 **(1)**

(b) disproportionation because mercury is simultaneously oxidised and reduced **(1)**

6 Bromide ions act as reducing agents **(1)** because they lose electrons **(1)**.

22. Ionic half-equations

1 (a) $Mg \rightarrow Mg^{2+} + 2e^-$ **(1)**

(b) $Cu^{2+} + 2e^- \rightarrow Cu$ **(1)**

(c) $Mg + Cu^{2+} \rightarrow Mg^{2+} + Cu$ **(1)**

2 (a) $Br_2 + 2I^- \rightarrow 2Br^- + I_2$ **(1)**

(b) $NO_3^- + V^{3+} \rightarrow NO_2 + VO^{2+}$
1 *mark for correct balancing* **1** *mark for cancelling out* H_2O *and* H^+

(c) $Cr_2O_7^{2-} + 3H_2O_2 + 8H^+ \rightarrow 3Cr^{3+} + 7H_2O + 3O_2$
1 *mark for correct balancing* **1** *mark for cancelling out* H^+

3 (a) $MnO_4^- + 8H^+ + \mathbf{5e^-} \rightarrow Mn^{2+} + \mathbf{4}H_2O$ **(1)**

(b) $MnO_4^- + 8H^+ + 5Fe^{2+} \rightarrow Mn^{2+} + 4H_2O + 5Fe^{3+}$
1 *mark for correct factor of 5 for* $Fe^{2+} \rightarrow Fe^{3+}$ *half-equation* **1** *mark for the rest of the equation being correct*

4 $Ag^+(aq) + Cl^-(aq) \rightarrow AgCl(s)$
1 *mark for correct species and balancing* **1** *mark for correct state symbols*

23. Reactions of Group 2 elements

1 **C (1)**

2 (a) (Going down Group 2, the first ionisation energy) decreases **(1)**

(b) atomic radius increases outer electrons further from nucleus **(1)**
(nuclear charge increases but) shielding increases **(1)**

3 (a) $Ba(s) + 2H_2O(l) \rightarrow Ba(OH)_2(aq) + H_2(g)$
1 *mark for correct species and balanced,* **1** *mark for correct state symbols*

(b) effervescence / bubbling / fizzing **(1)**
colourless solution formed **(1)**

4 (a) $Ca(s) + 2Cl_2 \rightarrow CaCl_2$ **(1)**

(b) Calcium is oxidised because its oxidation number changes from 0 to +2. **(1)**
Chlorine is reduced because its oxidation number changes from 0 to −1. **(1)**

5 (a) $2Ra + O_2 \rightarrow 2RaO$ **(1)**

(b) flame / coloured flame rapid reaction **(1)**
white solid is produced **(1)**

24. Reactions of Group 2 compounds

1 **A (1)**

2 (a) $BaO + H_2O \rightarrow Ba(OH)_2$ **(1)**

(b) (i) $2HCl + Ba(OH)_2 \rightarrow BaCl_2 + 2H_2O$ **(1)**
(ii) $H_2SO_4 + Ba(OH)_2 \rightarrow BaSO_4 + 2H_2O$ **(1)**

(c) barium sulfate **(1)**
white **(1)**
$Ba^{2+} + SO_4^{2-} \rightarrow BaSO_4$ **(1)**

3 (a) Magnesium hydroxide is sparingly soluble / not very soluble. **(1)**

(b) $2HCl + Mg(OH)_2 \rightarrow MgCl_2 + 2H_2O$ **(1)**

4 to prevent precipitates caused by sulfite / carbonate ions. **(1)**

5 Wet lime mortar will be alkaline / have a pH above 7 (so it will irritate or damage skin). **(1)**
The idea of harm alone is not enough − a reason must be given.

25. Stability of carbonates and nitrates

1 A reaction in which heat is used to break down a reactant into or more products. **(1)**

2 **C (1)**

3 (a) $2KNO_3 \rightarrow 2KNO_2 + O_2$
1 mark for correct species. 1 mark for correct balancing.
(b) $2Ca(NO_3)_2 \rightarrow 2CaO + 4NO_2 + O_2$
1 mark for correct species. 1 mark for correct balancing.

4 B (1)

5 (a) $CaCO_3 \rightarrow CaO + CO_2$ (1)
(b) (i) $SrCO_3$ / strontium carbonate (1)
(ii) The strontium ion is larger / has a smaller charge density than the calcium ion. (1) The strontium ion affects the carbonate ion less / has less polarising power (than the calcium ion). (1) Mention of less energy needed to break C−O bond / polarisation of carbonate ion (1).
The opposite argument for calcium carbonate is acceptable.

26. Flame tests

1 (a) apparatus: nichrome wire / platinum wire / silica rod (1)
reagent: concentrated hydrochloric acid (1)
Dip wire or rod in acid then in the powder, then hold it in the edge of a blue Bunsen burner flame. (1)
(b) no flame colour for magnesium ions (1)

2 C (1)

3 *One mark for each correct box:*

Group 1 cation	Flame colour
K⁺	**lilac (1)**
Rb⁺ (1)	red-violet
Li⁺	**red (1)**
Cs⁺ (1)	blue
Na⁺	**yellow-orange / yellow (1)**

4 (a) strontium: red
calcium: yellow-red / brick-red (1)
(b) idea that both flame colours are red (1)

5 D (1)

27. Properties of Group 7 elements

1 B (1)

2 (a) (Going down Group 7, the boiling temperature increases) (1)
(b) Size of the molecules increases / M_r increases / number of electrons (per molecule) increases. (1)
Strength of the intermolecular forces increases. (1)
More energy needed to break intermolecular forces / to separate molecules. (1)
mention of London forces (1)

3 1 mark for each correct box, to 5 marks:

Group 7 element	State at room temperature	Colour of vapour
chlorine	**gas (1)**	yellow-green
bromine	**liquid (1)**	**orange-brown (1)**
iodine	**solid (1)**	**purple (1)**

4 (a) fluorine (1)
(b) Fluorine atoms have a lower nuclear charge / fewer protons (1)
but fluorine atoms have less shielding due to few completed electron shells. (1)
The distance between the nucleus and bonding pair is less in fluorine molecules. (1)
The opposite argument for astatine is acceptable.

5 The idea that bromine vaporises easily / vapour pushes bromine out of the pipette. (1)

28. Reactions of Group 7 elements

1 B (1)

2 (a) caesium chloride (1)
(b) $2Cs + Cl_2 \rightarrow 2CsCl$ (1)
(c) caesium is oxidised because its oxidation number changes from 0 to +1. (1)
Chlorine is reduced because its oxidation number changes from 0 to −1. (1)

3 (a) (i) $Br_2 + 2KI \rightarrow 2KBr + I_2$ (1)
(ii) $Br_2 + 2I^- \rightarrow 2Br^- + I_2$ (1)
(iii) Two layers form. (1)
The **top** layer is purple. (1)
(iv) The colour of the top layer would be orange (due to bromine because no reaction). (1)
(b) A brown colour would appear on the paper / purple vapour given off. (1)
Fluorine is more reactive than iodine and will displace iodine from its compounds / fluorine is a stronger oxidising agent. (1)

29. Reactions of chlorine

1 A reaction in which an element in a single species is simultaneously oxidised and reduced. (1)

2 D (1)

3 Chlorine kills microorganisms / bacteria in the water (which may cause disease). (1)
'Germs' is not sufficient.

4 (a) $Cl_2 + 2OH^- \rightarrow Cl^- + ClO^- + H_2O$ (1)
(b) (i) sodium chlorate(I) (1)
(ii) bleach / household bleach / disinfectant (1)

5 (a) $NaBrO_3$ (1)
(b) $3Br_2 + 6NaOH \rightarrow 5NaBr + NaBrO_3 + 3H_2O$
1 mark for correct species, 1 mark for correct balancing
(c) disproportionation / redox (1)
Bromine is reduced to bromide, oxidation number changes from 0 to −1. (1)
Bromine is oxidised to bromate(V), oxidation number changes from 0 to +5. (1)

30. Halides as reducing agents

1 a species that loses electrons / donates electrons (in reactions) (1)

2 A (1)

3 (a) All the oxidation numbers stay the same. (1)
(b) misty white fumes (1)
(c) (i) $2HBr + H_2SO_4 \rightarrow Br_2 + SO_2 + 2H_2O$ (1)
(ii) reducing agent (1)

4 (a) (i) $6I^- \rightarrow 3I_2 + 6e^-$ / $2I^- \rightarrow I_2 + 2e^-$ (1)
(ii) $H_2SO_4 + 6H^+ + 6e^- \rightarrow 4H_2O + S$ (1)
(b) oxidising agent (1)
(c) hydrogen sulfide (1)

31. Other reactions of halides

1 B (1)

2 (a) to prevent carbonate ions forming precipitates (1)
(b) (With NaF I would observe) no visible change / colourless solution (1)
(but with NaCl I would observe) a white precipitate. (1)

3 (a) cream (1)
(b) $KBr(aq) + AgNO_3(aq) \rightarrow KNO_3(aq) + AgBr(s)$
1 mark for equation, 1 mark for correct state symbols
(c) Add dilute ammonia, no change should be visible. (1)
Add concentrated ammonia, the precipitate should dissolve / disappear. (1)

4 (a) silver chloride: white (1) silver iodide: yellow (1)
(b) Add dilute ammonia to each solid (in a test tube). (1)
Silver chloride should dissolve / form a colourless solution. (1)
Silver iodide should not dissolve / remains as a solid. (1)

32. Exam skills 3

1 (a) (i) $Br_2 + 2I^- \rightarrow 2Br^- + I_2$ (1)
(ii) Bromine gains electrons (1) and is oxidised / acts as a reducing agent (1).
Iodide ions lose electrons (1) and are reduced / act as an oxidising agent (1).
(iii) A purple layer will form (1) on top of a colourless layer (1).
(b) (i) $3NaClO \rightarrow 2NaCl + NaClO_3$ (1)
(ii) The oxidation number of chlorine changes from +1 in NaClO (1)
to −1 in NaCl (1) and +5 in $NaClO_3$ (1).

(c) (i) $SO_4^{2-} + 8H^+ + 6e- \rightarrow S + 4H_2O$ /
$H_2SO_4 + 6H^+ + 6e^- \rightarrow S + 4H_2O$
1 mark for correct species, **1** mark for balancing
(ii) Black solid: iodine / I_2 **(1)**
Toxic gas: hydrogen sulfide / H_2S **(1)**
Equation:
$8KI + 9H_2SO_4 \rightarrow 4I_2 + H_2S + 8KHSO_4 + 4H_2O$
1 mark for correct species, **1** mark for balancing
(d) Add dilute nitric acid **(1)** followed by dilute silver nitrate solution. **(1)**
A yellow precipitate forms **(1)** which is not soluble in concentrated ammonia solution **(1)**.

33. Moles and molar mass

1 (One mole is the amount of substance that contains) the same number of particles **(1)** as the number of carbon atoms **(1)** in exactly 12 g of ^{12}C. **(1)**
2 D **(1)**
3 (a) mol **(1)**　　　(b) $g\,mol^{-1}$ **(1)**　　　(c) no units **(1)**
4 (a) molar mass = $44.0\,g\,mol^{-1}$ **(1)**
mass = $1.0 \times 44.0 = 44.0\,g$ **(1)**
(b) molar mass = $58.5\,g\,mol^{-1}$ **(1)**
mass = $0.2 \times 58.5 = 11.7\,g$ **(1)**
(c) molar mass = $92.0\,g\,mol^{-1}$ **(1)**
mass = $2.5 \times 92.0 = 230\,g$ **(1)**
(d) molar mass = $32.0\,g\,mol^{-1}$ **(1)**
mass = $5.0 \times 10^{-3} \times 32.0 = 0.16\,g$ **(1)**
5 mass = $4.0 \times 2 \times 1.0$
= $8.0\,g$ **(1)**

34. Empirical and molecular formulae

1 The simplest whole number ratio **(1)** of the atoms of each element in a substance. **(1)**
2

Symbol	Cr	N	O	
Mass /g	0.874	0.706	2.42	
Molar mass	52.0	14.0	16.0	
g ÷ molar mass	$0.874 \div 52.0$ $= 0.0168$	$0.706 \div 14.0$ $= 0.0504$	$2.42 \div 16.0$ $= 0.151$	**(1)**
Divide by smallest value	$0.0168 \div$ $0.0168 = 1$	$0.0504 \div$ $0.0168 = 3$	$0.151 \div$ $0.0168 = 9$	**(1)**

Empirical formula is CrN_3O_9. **(1)**
3 mass of nitrogen = $0.505 - 0.365 = 0.140\,g$ **(1)**
Mg $0.365 \div 24.3 = 0.015$
N $0.140 \div 14.0 = 0.01$ **(1)**
Mg:N = 1.5:1 = 3:2
Empirical formula is Mg_3N_2. **(1)**
4 (a) $100 - 85.7 = 14.3\%$ **(1)**
(b) C $85.7 \div 12.0 = 7.14$　H $14.3 \div 1.0 = 14.3$ **(1)**
C:H = 1:2
Empirical formula is CH_2 **(1)**
(c) molar mass of $CH_2 = 14.0\,g\,mol^{-1}$
factor = $42.0 \div 14.0 = 3$ **(1)**
molecular formula is C_3H_6 **(1)**

35. Reacting masses calculations

1 molar mass of C = $12.0\,g\,mol^{-1}$
molar mass of $CO_2 = 44.0\,g\,mol^{-1}$
amount of carbon = $3.0 \div 12.0 = 0.25\,mol$ **(1)**
amount of carbon dioxide = $0.25\,mol$ **(1)**
mass of $CO_2 = 0.25 \times 44.0 = 11.0\,g$ **(1)**
2 molar mass of Mg = $24.3\,g\,mol^{-1}$
molar mass of $O_2 = 32.0\,g\,mol^{-1}$
amount of Mg = $1.50 \div 24.3 = 0.06173\,mol$ **(1)**
amount of $O_2 = 1 \div 2 \times 0.06173 = 0.03087\,mol$ **(1)**
mass of $O_2 = 0.03087 \times 32.0 = 0.988\,g$ **(1)** *must be to 3 significant figures*
3 molar mass of CO = $28.0\,g\,mol^{-1}$
molar mass of Fe = $55.8\,g\,mol^{-1}$
amount of CO = $3.00 \times 10^6 \div 28.0 = 1.07 \times 10^5\,mol$ **(1)**
amount of Fe = $2 \div 3 \times 1.07 \times 105 = 7.14 \times 10^4\,mol$ **(1)**
mass of Fe = $7.14 \times 10^4 \times 55.8 = 3.99 \times 10^6\,g$ / 3.99 tonnes **(1)**

4 (a) amount of iron = $2.79 \div 55.8 = 0.050\,mol$ and amount of carbon dioxide = $1.10 \div 44.0 = 0.025\,mol$ **(1)**
(b) Fe and CO_2 are in the ratio 0.050 : 0.025 = 2:1 **(1)** (so equation could be: $2FeO + C \rightarrow 2Fe + CO_2$)

36. Gas volume calculations

1 A **(1)**
2 (a) $2 \times 50 = 100\,cm^3$ **(1)**
(b) excess chlorine = $50\,cm^3$ **(1)**
volume of hydrogen chloride produced = $200\,cm^3$
total volume = $50 + 200 = 250\,cm^3$ **(1)**
3 (a) molar mass of butane = $58.0\,g\,mol^{-1}$ **(1)**
amount = $190 \div 58.0 = 3.28\,mol$ **(1)**
(b) amount of $O_2 = 6.5 \times 3.28 = 21.3\,mol$ **(1)**
(c) volume = $24.0 \times 21.3 = 511\,dm^3$ **(1)**
Allow $24.0 \times$ (answer b).
(d) Volume = $24.0 \times 3.28 \times 4 = 315\,dm^3$ **(1)**
Allow $24.0 \times$ (answer a).
(e) molar mass of water = $18.0\,g\,mol^{-1}$ **(1)**
mass = $5 \times 3.28 \times 18.0 = 295\,g$ **(1)**

37. Concentrations of solutions

1 B **(1)**
2 (a) (i) $2500 \div 1000 = 2.5\,dm^3$ **(1)**
(ii) $50 \div 1000 = 0.05\,dm^3 / 5 \times 10^{-2}\,dm^3$ **(1)**
(b) (i) $1.25 \times 1000 = 1250\,cm^3 / 1.25 \times 10^3\,cm^3$ **(1)**
(ii) $0.02 \times 1000 = 20\,cm^3$ **(1)**
3 (a) $2 \div 4 = 0.5\,mol\,dm^{-3}$ **(1)**
(b) volume = $0.1\,dm^3$
$0.25 \div 0.1 = 2.5\,mol\,dm^{-3}$ **(1)**
4 (a) amount = $0.5 \times 2 = 1.0\,mol$ **(1)**
(b) volume = $0.2\,dm^3$
amount = $0.2 \times 0.25 = 0.05\,mol$ **(1)**
5 (a) volume = $0.25\,dm^3$
amount = $0.25 \times 0.1 = 0.025\,mol$ **(1)**
(b) molar mass = $106.0\,g\,mol^{-1}$ **(1)**
(c) mass = $0.025 \times 106.0 = 2.65\,g$ **(1)**

38. Doing a titration

1 (a) It makes the colour change easier to see. **(1)**
(b) Phenolphthalein is a weak acid **(1)** so the titre of acid needed will be less than expected **(1)**.
(c) The readings will be higher than expected because the meniscus (at the front of the burette) will be below where it should be. **(1)**
(d) The titre will be greater than expected because some of the volume measured will be due to air not acid. **(1)**
2 (a) the point at which the indicator first changes colour **(1)**
(b) results within $0.10\,cm^3$ of each other / within a stated range **(1)**
3 $100 \times 0.06 \div 25 = 0.24\%$ **(1)**
4 $100 \times 0.10 \div 22.95 = 0.44\%$ **(1)**
Answer must be to 2 significant figures.

39. Titration calculations

1 (a) $29.20 \times 10^{-3} \times 0.100 = 2.92 \times 10^{-3}\,mol$ **(1)**
(b) $(2.92 \times 10^{-3}) \div 2 = 1.46 \times 10^{-3}\,mol$ **(1)**
(c) $10 \times 1.46 \times 10^{-3} = 1.46 \times 10^{-2}\,mol$ **(1)**
(d) molar mass = $1.55 \div (1.46 \times 10^{-2}) = 106$ **(1)**
(e) molar mass of $CO_3 = 60.0\,g\,mol^{-1}$
molar mass of M = $(106 - 60.0) \div 2 = 23.0\,g\,mol^{-1}$ **(1)**
M_2CO_3 must be sodium carbonate **(1)**
2 B **(1)**
3 amount of ethanedioic acid
= $27.50 \times 10^{-3} \times 0.0500 = 1.375 \times 10^{-3}\,mol$ **(1)**
amount of sodium hydroxide
= $1.375 \times 10^{-3} \times 2 = 2.75 \times 10^{-3}\,mol$ **(1)**
concentration of sodium hydroxide
= $(2.75 \times 10^{-3}) \div (25.0 \times 10^{-3}) = 0.110\,mol\,dm^{-3}$ **(1)**

40. Atom economy

1 percentage yield = $1.6 \div 2.0 \times 100 = 80\%$ **(1)**
2 (a) molar masses in $g\,mol^{-1}$ (all three for **1** mark):
CuO = 79.5　　CuSO_4 = 159.5　　H_2O = 18.0

(b) theoretical yield = 2.5 ÷ 79.5 × 159.5 = 5.0 g **(1)**

(c) percentage yield = 3.5 ÷ 5.0 × 100 = 70% **(1)**

3 D **(1)**

4 (a) total molar mass of product = 2 × 2.0 = 4.0 g mol⁻¹

total molar mass of all products = 4.0 + (2 × 16.0) = 36.0 g mol⁻¹

*Both for **1** mark*

atom economy = 4.0 ÷ 36.0 × 100 = 11.1% **(1)**

(b) sell the oxygen / use the oxygen in another process **(1)**

5 C **(1)**

41. Exam skills 4

1 (a) $25.40 \times 10^{-3} \times 0.100 = 2.54 \times 10^{-3}$ mol **(1)**

(b) $10 \times 2.54 \times 10^{-3} = 0.0254$ mol **(1)**

(c) $75.0 \times 10^{-3} \times 0.500 = 0.0375$ mol **(1)**

(d) excess acid = $0.0375 - 0.0254 = 0.0121$ mol **(1)**

amount of $NaHCO_3$ = 0.0121 mol **(1)**

(e) molar mass of $NaHCO_3$ = 100.1 g mol⁻¹ **(1)**

mass of $NaHCO_3$ = 100.1 × 0.0121 = 1.211 g **(1)**

percentage = (1.211 / 1.48) × 100 **(1)** = 81.8% **(1)**

must be three significant figures

(f) amount of CO_2 = 0.0121 mol **(1)**

volume of CO_2 = 0.0121 × 24 = 0.290 dm³ **(1)**

must be three significant figures (allow 290 cm³)

42. Alkanes

1 D **(1)**

2 Saturated: only single bonds / no C=C bonds / cannot undergo addition reactions. **(1)**

Hydrocarbon: only contains hydrogen and carbon. **(1)**

3 (a) C_4H_{10} **(1)**   (b) C_2H_5 **(1)**

(c) Displayed formula **(1)**   Skeletal formula **(1)**

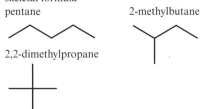

4 (a) $C_{12}H_{26}$ **(1)**

(b) $CH_3CH_2CH_2CH_2CH_2CH_2CH_2CH_2CH_2CH_2CH_2CH_3$ / $CH_3(CH_2)_{10}CH_3$ **(1)**

5 Pentane **(1)** C_5H_{12} **(1)**

43. Isomers of alkanes

1 C **(1)**

2 Name: 2,2-dimethylbutane **(1)**

Structural formula: $CH_3CH_2C(CH_3)_3$ / $(CH_3)_3CCH_2CH_3$ **(1)**

3 (a) **1** mark for each correct name, **1** mark for each correct skeletal formula

pentane   2-methylbutane

2,2-dimethylpropane

(b) structural isomerism / chain isomerism **(1)**

4 B **(1)**

5 (a) 2,2,3-trimethylbutane **(1)**   (b) 3-ethylpentane **(1)**

44. Alkenes

1 B **(1)**

2 (a) It contains C=C bonds / will undergo addition reactions. **(1)**

(b) Add bromine / bromine water. **(1)**

No colour change (stays orange / brown) in hexane but is decolourised in hex-1-ene. **(1)**

3 (a) $CH_3CH=CH_2$ / $CH_2=CHCH_3$ **(1)**

(b) Displayed formula **(1)**

Skeletal formula **(1)**

4 *1 mark for each correctly completed box to **6** marks maximum.*

	Saturated hydrocarbon	Unsaturated hydrocarbon
Displayed formula		
Skeletal formula		
Name	cyclopropane	propene

45. Isomers of alkenes

1 D **(1)**

2 (a) The C=C bond restricts rotation / C=C bond cannot rotate. **(1)**

The groups cannot change position (relative to the C=C bond). **(1)**

(b) (i) **1** mark for each correct structure:

E-but-2-ene   *Z*-but-2-ene

(ii) *E*-but-2-ene: *trans*-but-2-ene

Z-but-2-ene: *cis*-but-2-ene

*Both needed for **1** mark.*

(c) (i) 2-methylprop-1-ene **(1)**

(ii) One of the C atoms in the C=C has two identical groups (H atoms on but-1-ene and also methyl groups on 2-methylprop-1-ene). **(1)**

3 C **(1)**

46. Using crude oil

1 (a) boiling point / chain length **(1)**

(b) mixture of compounds with similar boiling temperatures / chain length / number of carbon atoms **(1)**

2 B **(1)**

3 (a) $C_6H_{14} \rightarrow C_4H_{10} + C_2H_4$ / $C_6H_{14} \rightarrow C_2H_6 + 2C_2H_4$ **(1)**

(b) To match the supply of smaller hydrocarbons to demand for them / Smaller hydrocarbons are more valuable as fuels. **(1)**

4 (a) 2-methylpentane **(1)**   (b) hexane **(1)** C_6H_{14} **(1)**

(c) (i) cyclohexane **(1)** C_6H_{12} **(1)**

(ii) hydrogen/H_2 **(1)**

47. Hydrocarbons as fuels

1 (a) $CH_4 + 2O_2 \rightarrow CO_2 + 2H_2O$ **(1)**

(b) enhanced greenhouse effect / global warming / climate change **(1)**

Carbon dioxide is a greenhouse gas. **(1)**

2 (a) If there is insufficient air (or oxygen) / a limited supply of air (or oxygen). **(1)**

(b) $C_6H_{14} + 5O_2 \rightarrow 3CO + 3C + 7H_2O$

1 mark for correct species, 1 mark for balancing.

(c) Any two for **1** mark each:

• Carbon monoxide is toxic.

• Carbon particulates / smoke / soot cause breathing problems.

• Unburned hydrocarbons are harmful / toxic.

• Less energy is released (than with complete combustion).

3 A **(1)**

4 Acid rain **(1)** because these oxides have acidic properties / dissolve in clouds to form acidic solutions. **(1)** Identified problem caused by acid rain, e.g. damage to buildings / stone / rivers / trees / aquatic life **(1)**.

5 (a) $2CO + 2NO \rightarrow 2CO_2 + N_2$ **(1)**   (b) redox **(1)**

48. Alternative fuels

1 C **(1)**

2 (An activity that has no) overall / net emissions of carbon / carbon dioxide (to the atmosphere). **(1)**

3 (a) Fuel derived from biomass / not derived from fossil fuels / can be replaced. **(1)**

(b) Fuels which cannot be replaced once they have all been used up / take a very long time to make and are used up faster than they are formed. **(1)**

Note that the word 're-used' is not acceptable instead of 'replaced'.

4 (a) process 1: plant sugars / sugar cane / sugar beet / molasses **(1)**
Sugar or a named sugar = 0 marks
process 2: crude oil **(1)**

(b) Two sensible advantages of fermentation over direct hydration for **1** mark each, e.g.
- Raw material is renewable.
- Product is carbon neutral / closer to being carbon neutral.
- Process uses a lower temperature / has lower energy costs.
- Process uses a lower pressure / cost of machinery is lower.

(c) Two sensible disadvantages of fermentation over direct hydration for **1** mark each, e.g.
- Process is slower.
- Product must be purified.
- Lower yield
- Lower atom economy
- More labour intensive
- Farmland is used which could be used to grow food.

5 Rapeseed / sunflower / soya / palm oil **(1)**

49. Alcohols and halogenoalkanes

1 Three of the following for **1** mark each:
- Same functional group
- Same general formula
- Similar chemical properties
- Trend in physical properties / named property, e.g. boiling temperature
- Successive members differ by CH_2

2 (a) (i) methanol **(1)**
(ii) propan-1-ol **(1)**
(iii) 2-methylbutan-2-ol **(1)**

(b) (ii) molecular formula: C_3H_8O **(1)**
structural formula: $CH_3CH_2CH_2OH$ **(1)**
(iii) molecular formula: $C_5H_{12}O$ **(1)**
structural formula: $(CH_3)_2CH(OH)CH_2CH_3$ **(1)**

(c) hydroxyl **(1)**

3 B **(1)**

4 **1** mark for each correct structure:

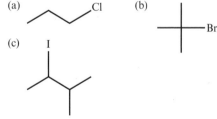

50. Substitution reactions of alkanes

1 A **(1)**

2 (a) (free) radical substitution **(1)**
(b) (i) $Br_2 \rightarrow \bullet Br + \bullet Br$ / $Br_2 \rightarrow 2Br\bullet$ **(1)**
(ii) $CH_4 + \bullet Br \rightarrow \bullet CH_3 + HBr$ **(1)**
(iii) $\bullet CH_3 + Br_2 \rightarrow CH_3Br + \bullet Br$ **(1)**
(iv) Two radicals combining to form a molecule, e.g.
$\bullet CH_3 + \bullet Br \rightarrow CH_3Br$ / $\bullet CH_3 + \bullet CH_3 \rightarrow C_2H_6$ **(1)**

(c) fractional distillation **(1)**

3 (a) CH_3Cl − chloromethane **(1)**
CH_2Cl_2 − dichloromethane **(1)**
(b) $CH_3Cl + \bullet Cl \rightarrow \bullet CH_2Cl + HCl$ **(1)**
$\bullet CH_2Cl + Cl_2 \rightarrow CH_2Cl_2 + \bullet Cl$ **(1)**
(c) $\bullet CH_2Cl + \bullet CH_3 \rightarrow CH_3CH_2Cl$ /
$\bullet CH_2CH_3 + \bullet Cl \rightarrow CH_3CH_2Cl$ **(1)**

4 A species with an unpaired electron **(1)**

51. Alkenes and hydrogen halides

1 C **(1)**

2 (a) Overlap of s-orbitals in σ bond is good because it is end-on / along the line between the two nuclei. **(1)**
Overlap of p-orbitals in π bond is poor because it is sideways / parallel to the line between the two nuclei. **(1)**

(b) *1 mark for each labelled bond to 2 marks, e.g.*

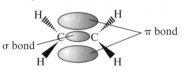

Sigma bond must not just be shown as a straight line.

3 (An electrophile is a species that can) accept / gain a <u>pair</u> of electrons. **(1)**

4 (a) Right hand box ticked **(1)**
(b) Diagram completed to show:
- curly arrow from C=C bond to H in Br **(1)**
- curly arrow from H−Br bond to Br in HBr **(1)**
- curly arrow from lone pair of electrons in Br^- ion to + on centre C atom of the intermediate **(1)**
- Correct product **(1)**

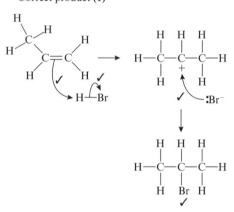

The carbocation involved has already been identified in part (a) but still needs to be completed in the diagram.

52. More addition reactions of alkenes

1 (a) (i) secondary **(1)** (ii) primary **(1)** (iii) tertiary **(1)**
(b) species (iii) **(1)**

2 (a) catalyst **(1)**
(b) with hydrogen: butane **(1)**
with bromine: 1,2-dibromobutane **(1)**
(c) Dipole is induced as the molecule approaches the C=C bond in the alkene. **(1)**
Electrons in the π bond repel the electrons in the H−H or Br−Br bond. **(1)**

3 (a) 1-bromobutane **(1)**
(b) Diagram completed to show:
- curly arrow from C=C bond to H in Br **(1)**
- curly arrow from H−Br bond to Br in HBr **(1)**
- curly arrow from lone pair of electrons in Br^- ion to + on second C atom of the intermediate **(1)**
- correct carbocation **(1)**
- correct product **(1)**

53. Exam skills 5

1 (a) (i) chlorine / Cl_2 **(1)** ultraviolet light /
450 °C−1000 °C **(1)**

(ii) $Cl_2 \rightarrow \bullet Cl + \bullet Cl$ **(1)**

(iii) $CH_3CH_2CH_3 + \bullet Cl \rightarrow CH_3CH_2\overset{\bullet}{C}H_2 + HCl$ **(1)**
$CH_3CH_2\overset{\bullet}{C}H_2 + Cl_2 \rightarrow CH_3CH_2CH_2Cl + \bullet Cl$ **(1)**

(iv) Two propyl radicals may react together to form hexane in an termination step **(1)**
$2CH_3CH_2\overset{\bullet}{C}H_2 \rightarrow C_6H_{14}$ **(1)**

(v) Use fractional distillation **(1)** because the different substances will have different boiling points **(1)**.

(b) (i)

1 mark for curly arrow from C=C bond to H and H−Cl bond to Cl. **1** mark for correct carbocation.
1 mark for curly arrow from chloride ion to + charge on carbocation.

(ii) electrophilic addition **(1)**

(iii) 1-chloropropane forms from a primary carbocation **(1)**
which is less stable than a secondary carbocation (which leads to the major product) **(1)**

54. Alkenes and alcohols

1 (a) $C_2H_4 + H_2O \rightarrow C_2H_5OH$ /
$CH_2=CH_2 + H_2O \rightarrow CH_3CH_2OH$ **(1)**

(b) type: hydration **(1)**
mechanism: electrophilic addition **(1)**

(c) elimination / dehydration **(1)**

(d) It is a catalyst. **(1)**

2 (a) B **(1)**

(b) (i) propane-1,2-diol **(1)**

(ii) $CH_3CH=CH_2 + [O] + H_2O \rightarrow CH_3CH_2(OH)CH_2OH$
1 mark for correct reactants.
1 mark for correct product (and balanced).

(iii) It is an oxidising agent / oxidant. **(1)**

(c) sulfuric acid **(1)**

55. Addition polymerisation

1 (a) repeat unit **(1)**   monomer **(1)**

Cl can be shown above or below.   *Cl can be shown above or below*

(b) monomer: chloroethene **(1)**
polymer: poly(chloroethene) **(1)**

2 (a) repeat unit **(1)**   section of polymer **(1)**

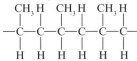

Diagram should show a minimum of two repeat units (there are three here).

(b) *Any one of the following for **1** mark:*
- Both have the same C : H **ratio** (*not the same number of C and H atoms*).
- No atoms are lost / gained.
- Polymer is produced by addition / no other product formed / one product formed.

(c) 100% because: only one product / no waste product / no other product / addition reaction **(1)**

3 Ethene decolourises bromine water but no change (stays orange) with poly(ethene). **(1)**
Ethene is unsaturated / contains C=C bonds, but poly(ethene) is saturated / has no C=C bonds. **(1)**

56. Polymer waste

1 (a) burning (at high temperatures) **(1)**
to release energy / dispose of waste **(1)**

(b) (i) $CaCO_3 + SO_2 \rightarrow CaSO_3 + CO_2$ **(1)**

(ii) acid rain / breathing difficulties / toxic gas / acidic gas **(1)**

2 (a) can be broken down / decomposed **(1)**
by bacteria / living organisms **(1)**

(b) The lactic acid monomer can be used to make poly(lactic acid) again / cracking does not produce the original monomer. **(1)**
The process uses less energy / lower temperatures / fewer resources than cracking. **(1)**

3 (a) Advantage: fewer raw materials needed / less energy needed / bag is lightweight **(1)**
Disadvantage: if used once, disposal problem / litter problem / waste of resources **(1)**

(b) Simple comparison for **1** mark, e.g. 'Bag for life' uses more raw materials and energy.
Quantitative comparisons for **1** mark each, e.g. 'Bag for life' uses 4 times more raw materials / 8 times more energy. (**2** marks maximum for above points)
Sensible conclusion for **1** mark, e.g. more sustainable if reused at least eight times / must be re-used at least eight times to equal the extra energy (and raw materials) used to make it / heavier bag so uses more energy to transport it.

57. Alcohols from halogenoalkanes

1 D **(1)**

2 (a) 1-bromopropane **(1)**   (b) propan-1-ol **(1)**

(c) 1-chloro-2-methylpropane **(1)**

(d) 3-methylbutan-1-ol **(1)**

3 (A nucleophile is a species that can) donate a lone pair of electrons. **(1)**

4 Reflux allows a reaction mixture to be heated for a long time without losing any liquid. **(1)**
The reaction mixture is heated and its vapours rise (into the condenser). **(1)**
The vapours are cooled and condensed, and liquid falls back into the reaction mixture. **(1)**

5 (a) correct structure of 1-bromobutane **(1)**
correct structure of butan-1-ol **(1)**
bromide ion shown as the inorganic product : Br^- **(1)**
(**1** mark for each correct curly arrow to two marks maximum)

(b) K^+ ion − spectator ion **(1)** OH^- ion − nucleophile **(1)**

58. Reactivity of halogenoalkanes

1 (a) (i) secondary **(1)**   (ii) tertiary **(1)**   (iii) primary **(1)**

(b) compound (ii) / the tertiary one **(1)**

2 breaking down of a molecule using water **(1)**

3 (a) $CH_3CH_2CH_2Br + H_2O \rightarrow CH_3CH_2CH_2OH + HBr$ **(1)**
You can show $H^+ + Br^-$ instead of HBr.
propan-1-ol **(1)**

(b) nucleophilic substitution **(1)**
*Both words needed for **1** mark.*

4 (a) Cl^- – white, Br^- – cream, I^- – yellow

All three colours correct for 1 mark.

(b) It is a solvent / it prevents two layers forming. **(1)**

5 Chloroethane is the least reactive and iodoethane is the most reactive. **(1)**

Explanation includes these points:
- The carbon–halogen bond is broken in the reaction. **(1)**
- The C–Cl bond is the strongest and the C–I bond is the weakest. **(1)**
- The lower the bond enthalpy, the more easily the bond is broken and the more reactive the halogenoalkane. **(1)**

59. More halogenoalkane reactions

1 **D (1)**

You will see all three other names in use; propan-1-amine is the IUPAC recommended name.

2 (a) **1** mark for the correct structure for the positively charged intermediate

The + sign must be placed next to the nitrogen atom.

1 mark for each correct curly arrow to **4** marks maximum

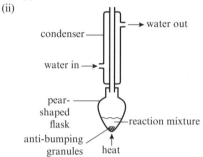

The curly arrow can come from any of the three N–H bonds, but the curly arrow from the :NH_3 molecule must point to the H atom involved.

(b) (In the first step, NH_3 acts as) a nucleophile. **(1)**

(In the second step, NH_3 acts as) a base. **(1)**

3 (a) $CH_3CH_2CH_2Br + KCN \rightarrow CH_3CH_2CH_2CN + KBr$ **(1)**

butanenitrile **(1)**

(b) cyanide ion / CN^- ion **(1)**

4 (a) $CH_3CHBrCH_3 + KOH \rightarrow CH_3CH_2=CH_2 + H_2O + KBr$ **(1)**

propene **(1)**

(b) type of reaction: elimination **(1)**

role of OH^- ion: base **(1)**

60. Oxidation of alcohols

1 (a) tertiary **(1)**   (b) secondary **(1)**   (c) primary **(1)**

2 (a) $CH_3CH_2OH + 3O_2 \rightarrow 2CO_2 + 3H_2O$ **(1)**

(b) $CH_3CH_2OH + 1\frac{1}{2}O_2 \rightarrow CO + C + 3H_2O$ **(1)**

3 (a) equation: $CH_3CH_2OH + [O] \rightarrow CH_3CHO + H_2O$ **(1)**

name of organic product: ethanal **(1)**

(b) equation: $CH_3CH_2OH + 2[O] \rightarrow CH_3COOH + H_2O$ **(1)**

name of organic product: ethanoic acid **(1)**

4 (a) from propan-1-ol: propanal **(1)**

from propan-2-ol: propanone **(1)**

(b) (**1** mark for reagent to add, **1** mark for *both* correct observations), e.g.

Reagent	Observation with:	
	product from propan-1-ol	**product from propan-2-ol**
Benedict's solution / Fehling's solution	changes from blue solution to red precipitate	no change / stays blue
Tollens' reagent	silver mirror forms	no change / stays colourless

61. Halogenoalkanes from alcohols

1 (a) $CH_3CH_2CH_2OH + PCl_5 \rightarrow CH_3CH_2CH_2Cl + POCl_3 + HCl$

(**1** mark for correct reactants)

(**1** mark for correct products and balanced)

(b) (fractional) distillation / separating funnel **(1)**

(c) Measure the boiling temperature of the liquid using distillation apparatus. **(1)**

The closer this is to the accepted temperature, the purer the liquid. **(1)**

2 (a) potassium bromide in 50% concentrated sulfuric acid **(1)**

(b) $CH_3CH_2CH_2CH_2OH + HBr \rightarrow CH_3CH_2CH_2CH_2Br + H_2O$ **(1)**

3 (a) red phosphorus and iodine **(1)**

(b) $3CH_3CH_2CH_2OH + PI_3 \rightarrow 3CH_3CH_2CH_2I + H_3PO_3$

(**1** mark for correct species) (**1** mark for balancing)

4 (a) The substances are immiscible (and have different densities). **(1)**

(b) (anhydrous) sodium sulfate / calcium sulfate / magnesium sulfate / calcium chloride **(1)**

62. Exam skills 6

1 (a) (i) $CH_3CH_2CH_2OH + 2[O] \rightarrow CH_3CH_2COOH + H_2O$ **(1)**

(ii)

water out

condenser

water in

pear-shaped flask

anti-bumping granules

heat

reaction mixture

pear-shaped flask / round-bottomed flask with heat **(1)**

condenser set up for reflux **(1)** *the top of the condenser must be open*

water flowing in correct direction **(1)**

(iii) orange to green / blue **(1)**

(iv) They have high boiling temperatures. **(1)**

(v) Use fractional distillation / absorb the water using an anhydrous drying agent. **(1)**

(b) (i) $2P + 3I_2 \rightarrow 2PI_3$ **(1)**

(ii) $3CH_3CH_2CH_2OH + PI_3 \rightarrow 3CH_3CH_2CH_2I + H_3PO_4$ **(1)**

(iii) A yellow precipitate forms. **(1)**

63. Structures from mass spectra

1 (a) Peak X: 15, Peak Y: 43, Peak Z: 58

(*All three correct for 1 mark*)

(b) (Z is the molecular ion peak) so $M_r = 58$ **(1)**

(c) CH_3^+ **(1)**

+ sign is needed.

(d) (i) CH_3CO^+ **(1)**

+ sign is needed.

(ii) CH_3COCH_3 **(1)**

propanone **(1)**

2 **B (1)**

$m/z = 15$ is likely to be CH_3^+ (in both butane and methylpropane).

$m/z = 29$ is likely to be $CH_3CH_2^+$ (in butane, but not in methylpropane).

$m/z = 43$ is likely to be $CH_3CH_2CH_2^+$ (in butane) or $CH_3CHCH_3^+$ (in methylpropane).

$m/z = 58$ is due to the molecular ion in each case.

64. Infrared spectroscopy

1 **D (1)**

2 (a) **A (1)**

(b) O–H bonds are present because there is a peak at 3350 / $3750-3200 \, cm^{-1}$. **(1)**

C=O bonds are not present because there is **no peak** at $1740-1700 \, cm^{-1}$ / C=C bonds are not present because there is **no peak** at $1669-1645 \, cm^{-1}$. **(1)**

A carboxylic acid should have a C=O and an O–H bond, but an alcohol should only have an O–H bond. **(1)**

3 (a) $1669-1645 \, cm^{-1}$ due to C=C, because product does not have C=C bond **(1)**

(b) $3750-3200 \, cm^{-1}$ due to O–H, because product (ethanol) has O–H bond, but ethene does not **(1)**

4 (a) propanone **(1)**

(b) $1740-1700 \, cm^{-1}$ because product (propanone) has C=O bond, but propan-2-ol does not **(1)**

65. Enthalpy changes

1 **C (1)**

2 (a) temperature: specified, usually 298 K **(1)**
   pressure: 100 kPa / 100 000 Pa / 1×10^5 Pa **(1)**
   Not 1 atm / 1 bar / 101 325 Pa etc.

   (b) (This is the enthalpy change when) 1 mole of a
   substance / compound **(1)**
   is formed from its elements in their standard states **(1)**
   under 100 kPa pressure and specified temperature /
   298 K **(1)**.

   (c) $Na(s) + \frac{1}{2}Br_2(l) + 1\frac{1}{2}O_2(g) \rightarrow NaBrO_3(s)$
   1 mark for correct species and balanced
   1 mark for correct state symbols
   Multiples not allowed here, e.g. not $2Na(s) + Br_2(l) + 3O_2(g) \rightarrow 2NaBrO_3(s)$

3 **1 mark** for each correct row in the table:

Type of reaction	Temperature change	Sign of ΔH (+ or −)
Exothermic	increases / goes up	− / minus / negative
Endothermic	decreases / goes down	+ / plus / positive

4 Labelled enthalpy level diagram with these features:
   - vertical axis labelled only **(1)**
   - reactants higher than products, and labelled with their formulae **(1)**
   - upwards arrow labelled with the value for ΔH **(1)**

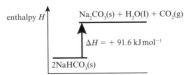

66. Measuring enthalpy changes

1 **D (1)**

2 (a) temperature change = (40.3 − 21.8) = 18.5 K **(1)**
   heat energy = $150 \times 4.18 \times 18.7 = 11\,600$ J **(1)**

   (b) molar mass of methanol = 32.0 g mol⁻¹ **(1)**
   mass of methanol burned = (123.64 − 122.78) = 0.86 g **(1)**
   amount of methanol = 0.86 ÷ 32.0 = 0.0269 mol **(1)**
   The final answer must be to three significant figures.

   (c) heat energy = 11 600 ÷ 1000 = 11.6 kJ **(1)**
   $\Delta_c H$ = 11.6 ÷ 0.0269 = −431 kJ mol⁻¹
   1 mark for negative sign and **1 mark** for answer (must be
   to three significant figures)

   (d) All the thermal energy from the burning fuel was used to
   heat up the water. **(1)**

3 The enthalpy change when 1 mole of water is produced **(1)**
   as a result of the reaction between an acid and an alkali
   / base **(1)** under standard conditions / 100 kPa and stated
   temperature / 298 K **(1)**.

67. Enthalpy cycles

1 (The enthalpy change for a reaction is) independent of the
   pathway taken **(1)** provided the initial and final conditions
   are the same for each pathway **(1)**.

2 **B (1)**

3 (a) $C(s) + 2H_2(g) + \frac{1}{2}O_2(g) \rightarrow CH_3OH(l)$ **(1)**
   Must include correct state symbols

   (b)
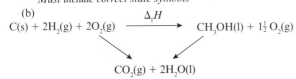

   1 mark for correct enthalpy cycle as shown
   Must not be multiplied through by 2 to remove fractions

4
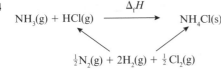

   1 mark for correct enthalpy cycle as shown
   Must not be multiplied through by 2 to remove fractions

68. Using enthalpy cycles

1 Using $\Delta_f H^\ominus$ data:
   $\Delta_r H^\ominus = (\Sigma \Delta_f H^\ominus[\text{products}]) - (\Sigma \Delta_f H^\ominus[\text{reactants}])$ **(1)**
   Using $\Delta_c H^\ominus$ data:
   $\Delta_r H^\ominus = (\Sigma \Delta_c H^\ominus[\text{reactants}]) - (\Sigma \Delta_c H^\ominus[\text{products}])$ **(1)**

2 Both enthalpy changes refer to $C(s) + O_2(g) \rightarrow CO_2(g)$ **(1)**

3 (a) The standard enthalpy change of formation for an
   element in its standard state is zero. **(1)**

   (b) $\Delta_c H^\ominus = (\Delta_f H^\ominus[CO_2(g)]) + 2\Delta_f H^\ominus[H_2O(l)]) - \Delta_f H^\ominus[CH_4(g)]$ **(1)**
   $\Delta_c H^\ominus = -394 + 2(-286) - (-75)$ kJ mol⁻¹ **(1)**
   $\Delta_c H^\ominus = -891$ kJ mol⁻¹ **(1)**

4 $\Delta_f H^\ominus = (2\Delta_c H^\ominus[C(s)]) + 3\Delta_c H^\ominus[H_2(g)]) - \Delta_c H^\ominus[CH_3CH_2OH(l)]$ **(1)**
   $\Delta_f H^\ominus = 2(-394) + 3(-286) - (-1367)$ kJ mol⁻¹ **(1)**
   $\Delta_f H^\ominus = -279$ kJ mol⁻¹ **(1)**

5 **C (1)**

69. Mean bond enthalpy calculations

1 (It is the enthalpy change when) one mole of a bond **(1)** in
   the **gaseous** state is broken **(1)**.

2 (a) Energy in to break bonds = 436 + (0.5 × 498) =
   685 kJ mol⁻¹ **(1)**
   Energy out when bonds form = 2 × 464 = 928 kJ mol⁻¹ **(1)**
   Enthalpy change = 685 − 928 = −243 kJ mol⁻¹ **(1)**

   (b) The bond enthalpy calculation is based on $H_2O(g)$, but
   $\Delta_c H^\ominus$ calculation is based on $H_2O(l)$ / energy released on
   changing from water vapour to liquid water **(1)**.

3 **A (1)**

4 (a) Energy in to break bonds: (4 × 413) + 612 + E(H−Cl) =
   2264 + E(H−Cl) kJ mol⁻¹ **(1)**
   Energy out when bonds form: (5 × 413) + 347 + 346 =
   2758 kJ mol⁻¹ **(1)**
   Enthalpy change: −97 = 2264 + E(H−Cl) − 2758 kJ mol⁻¹
   E(H−Cl): −97 − 2264 + 2758 = 397 kJ mol⁻¹ **(1)**

   (b) The actual bond enthalpies in the substances depend on
   the other atoms attached (to the bond), which may be
   different from the mean bond enthalpies used here. **(1)**

70. Changing reaction rate

1 **D (1)**

2 change in concentration of a substance (reactant or
   product) / amount of substance formed / used up / amount
   of product formed / amount of reactant used up **(1)**
   per unit time / in a given time **(1)**

3 (a) minimum amount of energy that colliding particles need
   for a reaction to happen / to initiate bond breaking in
   the reactants **(1)**

   (b) orientation of reactant molecules **(1)**

4 (a) The rate will increase. **(1)**

   (b) surface area increased **(1)** greater **frequency** of collisions
   between reactant particles **(1)**

5 (a) The rate will increase. **(1)**

   (b) more reactant particles in the same volume **(1)** greater
   frequency of collisions between reactant particles **(1)**

71. Maxwell−Boltzmann model

1 (a) **C (1)**

   (b) vertical line from peak of the curve, crossing the
   horizontal axis and labelled × **(1)**

   (c) the total number of gas molecules in the sample **(1)**

   (d) area shaded to the right of the E_a line and below curve
   T_1 **(1)**

   (e) T_2 curve:
   - is displaced to the right **and** lower than the T_1 curve **(1)**
   - starts at the origin, crosses the T_1 curve once and does
   not touch the horizontal axis **(1)**

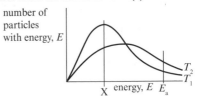

(f) most probable energy increases **(1)**, number of molecules with it decreases **(1)**

(g) increase in the number / fraction / proportion of molecules with the activation energy or more / more molecules have $E \geqslant E_a$ **(1)**
greater frequency of successful collisions / more successful collisions in a given time **(1)**

72. Catalysts

1 A **(1)**

2 (A catalyst provides) an alternative reaction route / pathway **(1)**
with a lower activation energy **(1)**.

3 enthalpy, H

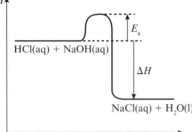

Correct shape of curve **(1)**
Reactants and products labelled **(1)**
E_a shown correctly **(1)**
$\Delta H^\ominus$ shown correctly **(1)**

4 enthalpy, H

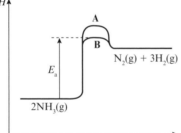

(a) Correct shape of curve and labelled **A (1)**
(b) Top of curve **B** lower than curve **a**; reactant and products levels unchanged **(1)**
(c) E_a shown correctly **(1)**

73. Dynamic equilibrium 1

1 The (forward and backward) reactions are still continuing. **(1)**

2 D **(1)**

3 They are the same / equal. **(1)**

4 (a) The equilibrium yield increases as the pressure increases. **(1)**
(b) There are more moles of gas on the left / fewer moles of gas on the right **(1)**,
decreasing the pressure moves the position of equilibrium in the direction of the most moles of gas **(1)**.
(c) The position of equilibrium will move to the right **(1)**,
because the rate of backward reaction is decreased / the rate of the forward reaction is increased **(1)**.

5 The position of equilibrium will move to the left **(1)**,
because the rate of the backward reaction is increased **(1)**.

74. Dynamic equilibrium 2

1 D **(1)**

2 (a) $-9.4\,\text{kJ}\,\text{mol}^{-1}$ **(1)**
(b) The position of equilibrium will move to the right / in the direction of the endothermic change **(1)**.
The rate of the endothermic reaction increases more than the rate of the exothermic reaction **(1)**.

3 (a) Correct expression for **1** mark:
$$K_c = \frac{[\text{C(aq)}]^c[\text{D(aq)}]^d}{[\text{A(aq)}]^a[\text{B(aq)}]^b}$$
(b) Correct expression for **1** mark:
$$K_c = \frac{[\text{SO}_3(g)]^2}{[\text{SO}_2(g)]^2[\text{O}_2(g)]}$$

4 (In a homogeneous system the components are) in the same phase / state, but in a heterogeneous system they are in two or more phases / states. **(1)**

5 It is constant / stays the same. **(1)**

6 (a) Correct expression for **1** mark:
$$K_c = \frac{[\text{SO}_3(g)]}{[\text{SO}_2(g)][\text{O}_2(g)]^{0.5}}$$
$\frac{1}{2}$ *could be used instead of 0.5*
(b) Correct expression for **1** mark:
$$K_c = \frac{[\text{Cu}^{2+}(ag)]}{[\text{Ag}^+(aq)]^2}$$
(c) Correct expression for **1** mark:
$$K_c = [\text{NH}_3(g)][\text{HCl}(g)]$$
(d) Correct expression for **1** mark:
$$K_c = \frac{[\text{HCl(aq)}][\text{HClO(aq)}]}{[\text{Cl}_2(aq)]}$$

75. Industrial processes

1 A **(1)**

2 Any **two** from the following for **1** mark each:
- higher cost of energy / electricity
- higher cost of equipment
- higher pressures more hazardous than lower pressures
- additional cost may not be met by the increased yield

3 (a) phosphoric acid / aluminium oxide **(1)**
(b) decreased yield **(1)**
(c) A high temperature gives a low yield but a low temperature gives a low rate **(1)**, this is a compromises temperature between yield and rate **(1)**.
(d) **Two** from the following for **1** mark each:
- increase the concentration of ethene / steam
- remove ethanol as it forms
- increase the pressure

4 Exothermic because:
the equilibrium yield decreases with increase in temperature **(1)** increasing temperature will increase the rate of the backward reaction / the endothermic reaction more **(1)**

76. Exam skills 7

1 (a) (i) Correct expression for **1** mark:
$$K_c = \frac{[\text{NO}_2(g)]^2}{[\text{N}_2(g)][\text{O}_2(g)]}$$
(ii) The equilibrium yield is increased **(1)** because the forward reaction is endothermic **(1)**.
(iii) There is no effect on the equilibrium yield if the pressure is increased **(1)** because the number of moles of gas is the same on both sides of the equation **(1)**.
(iv) There is no effect on the equilibrium yield if a catalyst is present **(1)** because the catalyst increases the rate of the forward reaction and backward reaction by the same ratio **(1)**.
(b) The equilibrium position moves to the left **(1)** because the rate of the backward reaction is increased **(1)**.
(c) (i) The reaction mixture is not allowed to reach equilibrium **(1)** as this would take too long / be uneconomical / overall rate of production is important **(1)**.
(ii) High pressures favour a high equilibrium yield of ammonia **(1)** because there are fewer moles of gas on the right of the equation **(1)** but very high pressures are expensive / need a lot of energy to produce / are not economical **(1)**.

77. Partial pressures and K_p

1 A **(1)**

2 (It is the pressure a gas would exert) on its own under the same conditions. **(1)**

3 (a) N_2 $3.2 / (3.2 + 1.8) = 0.64$ **(1)**
H_2 $1.8 / (3.2 + 1.8) = 0.36$ **(1)**
(b) $p_{\text{N}_2} = 0.64 \times 2.5 = 1.6\,\text{atm}$
$p_{\text{H}_2} = 0.36 \times 2.5 = 0.9\,\text{atm}$
Both needed for 1 mark

(c) 0.8 mol of NH_3 forms from:
  $(1 \times 0.8 / 2) = 0.4 \, mol \, N_2$
  $(3 \times 0.8 / 2) = 1.2 \, mol \, H_2$ **(1)**
  At equilibrium:
  $3.2 - 0.4 = 2.8 \, mol \, N_2$
  $1.8 - 1.2 = 0.6 \, mol \, H_2$ **(1)**

4 1 mark for each correct expression for K_p

(a) $K_p = \dfrac{p_{N_2O_4}}{(p_{NO_2})^2}$ (b) $K_p = \dfrac{p_{CO} \times p_{H_2}}{p_{H_2O}}$ (c) $K_p = \dfrac{p_{SO_3}}{p_{SO_2} \times (p_{O_2})^{0.5}}$

78. Calculating K_c and K_p values

1 D **(1)**

2 1 mark for correct expression for K_p

$K_p = \dfrac{p_{H_2} \times p_{CO_2}}{p_{H_2O} \times p_{CO}}$ **(1)**

Value for $K_p = (3.7 \times 0.6) / (0.2 \times 1.4) = 7.9$ **(1)** no units **(1)**

3 Units for K_c: $mol \, dm^{-3}$ **(1)** $mol^{-1} \, dm^3$ **(1)** $mol^2 \, dm^{-6}$ **(1)**
  Units for K_p: atm **(1)** atm^{-1} **(1)** atm^2 **(1)**

4 (a) amount of N_2O_4 reacted $= 0.120 - 0.045 = 0.075 \, mol$
  amount of NO_2 at equilibrium $= 2 \times 0.075 = 0.15 \, mol$ **(1)**

(b) $[N_2O_4] = 0.045 / 3.0 = 0.015 \, mol \, dm^{-3}$
  $[NO_2] = 0.15 / 3.0 = 0.05 \, mol \, dm^3$
  Both required for 1 mark

(c) *1 mark for correct expression for* K_c

$K_c = \dfrac{[NO_2]^2}{[N_2O_4]}$

$K_c = (0.05)^2 / 0.015$ **(1)**
  $= 0.167$ **(1)** $mol \, dm^{-3}$ **(1)**

79. Changing K_c and K_p

1 B **(1)**

2 C **(1)**

3 (a) 1 mark for correct expression for K_p

$K_p = \dfrac{(p_{NH_3})^2}{(p_{N_2})(p_{H_2})^3}$

(b) The catalyst increases the rate of the forward and backward reactions by the same ratio **(1)** so the position of equilibrium does not change / the equilibrium concentrations do not change / the equilibrium partial pressures do not change **(1)**.

(c) A catalyst increases the rate of the forward and backward reactions by the same ratio **(1)** so the position of equilibrium does not change **(1)**.

4 (a) endothermic **(1)**

(b) The value of K_p will increase **(1)** because the equilibrium position moves to the right / the equilibrium concentration of carbon dioxide / K_p depends on the concentration or pressure of carbon dioxide **(1)**.

80. Acids, bases and pH

1 Acid: proton donor **(1)** Base: proton acceptor **(1)**

2 C **(1)**

3 Hydrochloric acid can donate one proton **(1)** but sulfuric acid can donate two **(1)**.

4 (a) CN^- **(1)** (b) F^- **(1)**

5 HClO (acid) / ClO^- (conjugate base) **(1)**
  BrO^- (base) / HBrO (conjugate acid) **(1)**

6 (a) $pH = -\log_{10}[H^+]$ **(1)**
  (b) $pH = -\log_{10}(0.255) = 0.59$ **(1)**
  (c) $[H^+] = 10^{-2.8} = 1.58 \times 10^{-3} \, mol \, dm^{-3}$ **(1)**

81. pH of acids

1 (A strong acid is) fully dissociated in solution **(1)** (but a weak acid is) only partially dissociated in solution **(1)**.

2 $pH = -\log_{10}(0.0200) = 1.70$ **(1)** *answer must be to two decimal places*

3 (a) Correct expression for K_a for **1** mark,
  e.g. $K_a = \dfrac{[H^+(aq)][HCOO^-(aq)]}{[HCOOH(aq)]}$

(b) (i) $[H^+]^2 = K_a \times [HCOOH]$ **(1)**
  $[H^+] = \sqrt{1.60 \times 10^{-4} \times 0.275} = 6.633 \times 10^{-3}$ **(1)**
  $pH = -\log_{10}(6.633 \times 10^{-3}) = 2.18$ **(1)**

(ii) The degree of ionisation of the acid is negligible / very small. **(1)**
  All of the H^+ ions come from the acid / you can ignore the H^+ ions from water. **(1)**

4 B **(1)**

5 (a) $pK_a = -\log_{10}[K_a] = -\log_{10}(1.30 \times 10^{-3}) = 2.89$ **(1)**
  (b) $K_a = 10^{-pK_a} = 10^{-2.90} = 1.26 \times 10^{-3} \, mol \, dm^{-3}$ **(1)**

82. pH of bases

1 (a) $H_2O(l) \rightleftharpoons H^+(aq) + OH^-(aq)$ **(1)**
  (b) $K_w = [H^+(aq)][OH^-(aq)]$ **(1)**
  (c) $[H^+(aq)] = \sqrt{K_w} = 8.252 \times 10^{-8} \, mol \, dm^{-3}$ **(1)**
  $pH = -\log_{10}(8.252 \times 10^{-8}) = 7.08$ **(1)** *answer must be to two decimal places*
  (d) $pK_w = -\log_{10}(K_w) = -\log_{10}(1.47 \times 10^{-14}) = 13.8$ **(1)**

2 (a) It is fully dissociated in solution. **(1)**
  (b) $[OH^-] = 0.0125 \, mol \, dm^{-3}$ **(1)**
  $[H^+] = \dfrac{K_w}{[OH^-]} = \dfrac{(1.00 \times 10^{-14})}{0.0125} = 8.00 \times 10^{-13}$ **(1)**
  $pH = -\log_{10}[H^+] = -\log_{10}(8.00 \times 10^{-13}) = 12.1$ **(1)** *answer must be to three significant figures*

3 (a) Hydrochloric acid and nitric acid are both fully dissociated in solution. **(1)**
  (b) Hydrogen cyanide is a weak acid / partially dissociated in solution. **(1)**
  Energy is needed to dissociate it. **(1)**

83. Buffer solutions

1 a solution that resists changes in pH / maintains an **almost** constant pH **(1)**
  when **small** amounts of acid or base are added to it **(1)**

2 (a) $HCOOH(aq) \rightleftharpoons HCOO^-(aq) + H^+(aq)$ **(1)**
  (b) (Buffer contains) large concentrations of HCOOH and $HCOO^-$. **(1)**
  OH^- reacts with HCOOH / $HCOOH + OH^- \rightarrow HCOO^- + H_2O$ / OH^- reacts with H^+ to form H_2O and more HCOOH dissociates. **(1)**
  Ratio of HCOOH : $HCOO^-$ changes very little. **(1)**

3 D **(1)**

4 HCO_3^- **(1)**

5 (a) $[CH_3COOH] = 0.100 \times \dfrac{150}{200} = 0.075 \, mol \, dm^3$
  $[CH_3COONa] = 0.200 \times \dfrac{50}{200} = 0.050 \, mol \, dm^3$ *both needed for 1 mark*
  (b) Equation for calculating $[H^+]$ **(1)**, e.g.
  $[H^+] = K_a \times \dfrac{[CH_3COOH]}{[CH_3COO^-]}$
  $[H^+] = (1.74 \times 10^{-5}) \times \dfrac{0.075}{0.050} = 2.61 \times 10^{-5} \, mol \, dm^{-3}$ **(1)**
  (c) $pH = -\log_{10}(2.61 \times 10^{-5}) = 4.58$ **(1)**

84. More pH calculations

1 (a) $[H^+] = 10^{-pH} = 10^{-4.92} = 1.20 \times 10^{-5} \, mol \, dm^{-3}$ **(1)**
  (b) $[CH_3COONa] = K_a \times \dfrac{[CH_3COOH]}{[H^+]}$
  $= (1.74 \times 10^{-5}) \times \dfrac{0.500}{(1.20 \times 10^{-5})}$ **(1)** $= 0.725 \, mol \, dm^{-3}$ **(1)**
  (c) amount = concentration × volume $= 0.725 \times 200 \times 10^{-3}$
  $= 0.145 \, mol$ **(1)**
  (d) molar mass of $CH_3COONa = (2 \times 12.0) + (3 \times 1.0) + (2 \times 16.0) + (1 \times 23.0) = 82.0 \, g \, mol^{-1}$ **(1)**
  mass $= 0.145 \times 82.0 = 11.9 \, g$ **(1)** *three significant figures required*

2 (a) The pH of nitric acid increases by 1.0 for each dilution; the pH of ethanoic acid goes up by 0.5 for each dilution. **(1)**
  (b) *Nitric acid is a strong acid and ethanoic acid is a weak acid*
  Ethanoic acid is partially dissociated and it becomes more dissociated as it is diluted / the equilibrium $CH_3COOH \rightleftharpoons CH_3COO^-(aq) + H^+(aq)$ moves to the right. **(1)**
  The decrease in $[H^+]$ is less than for nitric acid / the $[H^+]$ is more than expected. **(1)**

85. Titration curves

1 (a) Curve A **(1)**   (b) Curve D **(1)**   (c) Curve B **(1)**
2 curve starts at in range pH 11–12 **(1)**
 vertical section at 25 cm³ with centre at about pH 4.9 (range 3–7) **(1)**
 end approaching pH 0.6 **(1)**
 shape **(1)**, e.g.

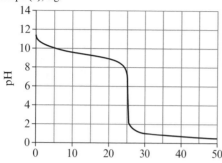

volume of hydrochloric acid added /cm³

86. Determining K_a

1 (a) The steep region in the titration curve is between about pH 7 and pH 3. **(1)**
  Methyl orange changes colour in this range but phenolphthalein does not. **(1)**
 (b) The pH change is too gradual near to the equivalence point / there is no vertical region in the titration curve. **(1)**
 (c) It forms a blue anion / negatively charged ion **(1)**
  because it loses H⁺ ions in alkaline solutions **(1)**.
2 (a) phenolphthalein **(1)**
  The equivalence point is about pH 8.5.
 (b) amount of NaOH = $0.100 \times 22.60 \times 10^{-3}$
   = 2.26×10^{-3} mol **(1)**
  amount of $CH_3COOH = 2.26 \times 10^{-3}$ mol **(1)**
  concentration of $CH_3COOH = \dfrac{(2.26 \times 10^{-3})}{(25.00 \times 10^{-3})}$
   = 0.0904 mol dm⁻³ **(1)**
 (c) at half-equivalence, pH = pK_a **(1)**
  pH = $-\log_{10}(1.74 \times 10^{-5})$ = 4.76 **(1)**
 (d) The flask contains ethanoic acid and sodium ethanoate / ethanoate ions. **(1)**
  These form a buffer solution (which resists changes in pH upon addition of alkali). **(1)**

87. Exam skills 8

1 (a) $(7 \times 12.0) + (6 \times 1.0) + (2 \times 16.0) = 122.0$ g mol⁻¹ **(1)**
 (b) (i) amount = 0.36 / 122.0 = 2.95×10^{-3} mol **(1)**
   concentration = $(2.95 \times 10^{-3}) / (250 \times 10^{-3})$
    = 1.18×10^{-2} mol dm⁻³ **(1)**
  (ii) There is negligible dissociation / ionisation of benzoic acid in solution. **(1)**
 (c) $[H^+] = 10^{-pH} = 10^{-3.06} = 8.71 \times 10^{-4}$ mol dm⁻³ **(1)**
 (d) **1** mark for correct expression, e.g.

  $$K_a = \dfrac{[C_6H_5COO^-(aq)][H^+(aq)]}{[C_6H_5COOH(aq)]}$$

 (e) (i) $K_a = (8.71 \times 10^{-4})^2 / (1.18 \times 10^{-2})$ **(1)**
    = 6.43×10^{-5} mol dm⁻³ **(1)**
  (ii) $[C_6H_5COO^-(aq)] = [H^+(aq)]$ **(1)**
  (iii) stated error in measurements, e.g. balance reads to ±0.01 g, pH meter reads to ±0.01, precision of volumetric flask, measurements not taken at 25 °C **(1)**

88. Born–Haber cycles 1

1 $Mg^{2+}(g) + 2Cl^-(g) \rightarrow MgCl_2(s)$ **(1)**
2 (a) (i) The enthalpy change when one mole of gaseous atoms **(1)** are formed from an element in its standard state **(1)**.
  (ii) $\frac{1}{2}Br_2(l) \rightarrow Br(g)$ **(1)** *must have correct state symbols*
 (b) (i) The enthalpy change when one mole of electrons is gained **(1)** by one mole of gaseous atoms (or anions) to form one mole of 1– charged gaseous anions **(1)**.
  (ii) $S^-(g) + e^- \rightarrow S^{2-}(g)$ **(1)** *must have correct state symbols*

3 (a) first ionisation energy of lithium **(1)**
 (b) **1** mark for each correctly placed symbol, e.g.

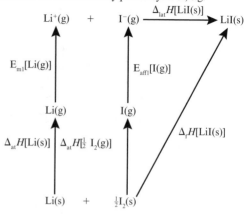

89. Born–Haber cycles 2

1 (a) Born–Haber cycle, e.g.

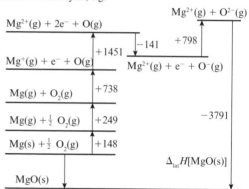

or

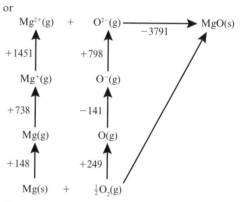

  (i) correct stage **(1)** all correct species **(1)** all correct state symbols **(1)**
  (ii) arrows in correct directions with values written next to them **(1)**
 (b) $\Delta_f H = 148 + 249 + 738 + 1451 + (-141) + 798 + (-3791)$ **(1)** = −548 kJ mol⁻¹ **(1)**
 (c) Electrons and O⁻ ions are both negatively charged. **(1)** There is repulsion between them which must be overcome. **(1)**

90. An ionic model

1 Ca^{2+} because
 it has the smallest ionic radius **(1)** and a 2+ charge (rather than 1+) **(1)**
 Allow greatest charge density with these reasons.
2 A **(1)**
3 (a) (Going down Group 7, the ionic radius) increases **(1)** the (electrostatic) force of attraction to the sodium ion in the lattice decreases **(1)** the lattice energy becomes less exothermic **(1)**.
 (b) Going down Group 1, the ionic radius increases **(1)** the (electrostatic) force of attraction to the bromide ion in the lattice decreases **(1)** the lattice energy becomes less exothermic **(1)**.

4 (a) Two from the following for **1** mark each:
  - ions are perfect spheres
  - charge on ions is distributed evenly
  - ions pack together in a regular arrangement.
  (b) AgI has covalent character **(1)** because the iodide ion is polarised / distorted by the silver ion **(1)**.

91. Dissolving

1 The enthalpy change when one mole of an ionic solid **(1)** dissolves in excess water to form an infinitely dilute solution **(1)**.

2 B **(1)**

3 bonds form between ions and water molecules **(1)** this is an exothermic process **(1)**

4 (a) energy cycle diagram, e.g.

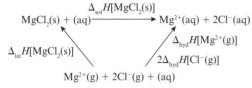

  1 mark for all species correct
  1 mark for arrows in correct directions and labelled
  (b) $\Delta_{sol}H = -\Delta_{lat}H[MgCl_2] + \Delta_{hyd}H[Mg^{2+}] + 2\Delta_{hyd}H[Cl^-]$
  $= -(-2493) + (-1920) + 2(-363)$ **(1)**
  $= -153\,kJ\,mol^{-1}$ **(1)**

5 The Mg^{2+} ion has a smaller ionic radius **(1)** and a higher charge **(1)** than the K^+ ion.

92. Entropy

1 increasing total entropy / positive entropy change **(1)**

2 D **(1)**

3 Entropy increases **(1)** because the ions are arranged more randomly in the solution than they are in the crystals **(1)**.

4 (a) gas **(1)**
  (b) randomly arranged **(1)**, move around each other **(1)**
  (c) There is a larger increase in entropy when water boils **(1)** some bonds break when water melts but more bonds break when it boils **(1)**, greater increase in movement when water boils **(1)** / greater increase in particle movement when water boils. **(1)**
  Accept reverse arguments.

5 C **(1)**

93. Calculating entropy changes

1 (a) It is an element in its standard state **(1)** so its standard enthalpy change of formation is zero **(1)**.
  (b) $\Delta S_{system}^{\ominus} = 197.6 + (3 \times 130.6) - 186.2 - 188.7$ **(1)**
  $= +214.5\,J\,K^{-1}\,mol^{-1}$ **(1)**
  (c) $\Delta H^{\ominus} = -110.5 - (-74.8) - (-241.8)$ **(1)**
  $= +206.1\,kJ\,mol^{-1}$ **(1)**
  (d) $\Delta S_{surroundings}^{\ominus} = 1000 \times (206.1)/298$ **(1)**
  $= +691.6\,J\,K^{-1}\,mol^{-1}$ **(1)**
  (e) $\Delta S_{total}^{\ominus} = 214.5 + 691.6 = +906.1\,J\,K^{-1}\,mol^{-1}$ **(1)**
  (f) The reaction is thermodynamically possible **(1)** ΔG is negative **(1)**.
  (g) $T = \Delta H^{\ominus}/\Delta S_{system}^{\ominus}$ **(1)** *because $\Delta G = 0$*
  $T = (206.1 \times 1000)/214.5$ **(1)**
  $T = 961\,K$ **(1)**

94. Gibbs energy and equilibrium

1 A **(1)**

2 (a) $\Delta G = (-9.6) - (298 \times 21.8 / 1000)$ **(1)**
  $= -16.1\,kJ\,mol^{-1}$ **(1)**
  (b) It is thermodynamically feasible **(1)** because ΔG is negative **(1)**.
  (c) (i) $\Delta G = -16.1 \times 1000 = -16\,100\,J\,mol^{-1}$ **(1)**
  $\ln K = -\Delta G/RT = 16\,100/(8.31 \times 298) = 6.50$ **(1)**
  (ii) $K = e^{6.50} = 665$ **(1)**
  (d) (i) K is much larger than 1 so the position of equilibrium lies far to the right **(1)** favouring a high equilibrium concentration of hydrogen iodide **(1)**.
  (ii) The activation energy must be very high **(1)**.

95. Redox and standard electrode potential

1 D **(1)**

2 (a) loss of electrons **(1)**
  (b) oxidation number of Cr is +3 in Cr^{3+} and +6 in CrO_4^{2-} **(1)**
  oxidation number of Cr increases in the reaction **(1)**

3 298 K temperature **(1)** 100 kPa pressure of gases **(1)**
  1.00 mol dm⁻³ concentration of ions **(1)**

4 Sketch of standard hydrogen electrode labelled to show:
  $H_2(g)$ **(1)** platinum electrode / platinum black electrode **(1)**
  $H^+(aq)$ / HCl(aq) **(1)**
  reasonable structure of the electrode, or conditions, e.g.
  100 kPa / 1.0 mol dm⁻³ / 298 K **(1)**

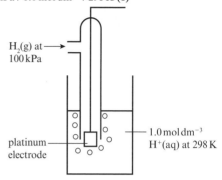

5 (a) (The value is) 0.00 V **(1)**
  (because) it is defined as this / it is the standard **(1)**.
  (b) Electrode potentials cannot be measured directly **(1)** they can only be compared against one another / to a standard **(1)**.

96. Measuring standard emf

1 A **(1)**

2 (a) Put magnesium (ribbon) into a solution containing Mg^{2+} ions / identified soluble magnesium salt, $MgCl_2(s)$, $Mg(NO_3)_2$. **(1)**
  (b) Put an inert electrode, e.g. platinum (*not* iron) **(1)** into a solution containing Fe^{2+} ions *and* Fe^{3+} ions / identified soluble iron(II) and iron(III) compounds **(1)**.

3 (a) KNO_3 / potassium nitrate **(1)** *not KCl because an AgCl precipitate would form*
  (b) allows ions to move between half-cells without mixing **(1)**
  (c) $Cu(s)|Cu^{2+}(aq) \colon\colon Ag+(aq)|Ag(s)$
  1 mark for correct use of | and ⋮⋮
  1 mark for species in correct order
  (d) $E_{cell}^{\ominus} = E_{right}^{\ominus} - E_{left}^{\ominus} = +0.80 - 0.34 = +0.46\,V$ **(1)**

97. Predicting reactions

1 B **(1)**

2 (a) $2Ag^+ + Fe \rightarrow 2Ag + Fe^{2+}$ **(1)**
  $E_{cell}^{\ominus} = +0.80 - (-0.44) = +1.24\,V$ **(1)**
  (b) $Ag^+ + Fe^{2+} \rightarrow Ag + Fe^{3+}$ **(1)**
  $E_{cell}^{\ominus} = +0.80 - (+0.77) = +0.03\,V$ **(1)**
  (c) The final species is Fe^{3+} **(1)** because Ag^+ ions can reduce Fe to Fe^{2+} *and then* Fe^{2+} to Fe^{3+} **(1)**.

3 (a) $E_{cell}^{\ominus}$ for feasible reaction $= +0.80 - (+0.54) = +0.26\,V$ **(1)**
  This is with Ag^+ being reduced to Ag and I^- oxidised to I_2, so silver will not react with iodine **(1)**.
  (b) $E_{cell}^{\ominus}$ for feasible reaction $= +0.77 - (-0.44) = +1.21\,V$ **(1)**
  This is with $Fe^{3+} + e^- \rightarrow Fe^{2+}$ (not $Fe^{2+} \rightarrow Fe^{3+} + e^-$) so Fe^{2+} cannot disproportionate **(1)**.

98. Limitations to predictions

1 D **(1)**

2 (a) $Cu + 2NO_3^- + 4H^+ \rightarrow Cu^{2+} + 2NO_2 + 2H_2O$ **(1)**
  (b) $E_{cell}^{\ominus}$ for feasible reaction $= +0.80 - (+0.34) = +0.46\,V$ **(1)**
  This value is positive so the reaction is feasible **(1)**.
  (c) Standard electrode potentials are for standard conditions **(1)** of 1 mol dm⁻³ / concentrated nitric acid is not standard conditions **(1)**.

3 (a) $Cl_2 + H_2O \rightarrow \frac{1}{2}O_2 + 2HCl$ /
  $Cl_2 + H_2O \rightarrow \frac{1}{2}O_2 + 2H^+ + 2Cl^-$ **(1)**
  $E_{cell}^{\ominus}$ for this reaction $= +1.36 - (+1.23) = +0.13\,V$ *with calculation E is positive so reaction is feasible* **(1)**.

(b) Light is needed to provide the activation energy / to break the Cl−Cl bond (by homolytic fission) **(1)**.

4 E_{cell} will increase **(1)**.
The position of the equilibrium $Zn^{2+}(aq) + 2e^- \rightleftharpoons Zn(s)$ moves to the right **(1)**
fewer electrons are released / right-hand electrode becomes more positive **(1)**.

99. Storage cells and fuel cells

1 (a) It develops a potential difference / voltage continuously while fuel is supplied **(1)**.
(b) (i) $E_{\text{cell}}^{\ominus} = +1.23 - (0.00) = +1.23\,\text{V}$ **(1)**
Must show working out
(ii) $H_2 + \frac{1}{2}O_2 \rightarrow H_2O$ **(1)**
(c) (i) $E_{\text{cell}}^{\ominus} = +0.40 - (-0.83) = +1.23\,\text{V}$ **(1)**
Must show working out
(ii) $2OH^- + H_2 + \frac{1}{2}O_2 + H_2O \rightarrow 2OH^- + 2H_2O$ **(1)**
$H_2 + \frac{1}{2}O_2 \rightarrow H_2O$ **(1)**
(d) The overall equations are the same **(1)** so the $E_{\text{cell}}^{\ominus}$ values are the same **(1)**.
2 (a) $E_{\text{cell}}^{\ominus} = +0.48 - (-0.81) = +1.29\,\text{V}$ **(1)**
(b) conditions are not standard / current is not negligible / voltmeter is not high-resistance **(1)**
(c) $2Ni(OH)_2 + Cd(OH)_2 \rightarrow 2Ni(OH)_3 + Cd$
1 mark for correction direction
1 mark for balancing with electrons and OH⁻ ions cancelled

100. Exam skills 9

1 (a) (i) $C_2H_5OH + 3O_2 \rightarrow 2CO_2 + 3H_2O$ **(1)**
(ii) $\frac{1}{2}O_2 + 2H^+ + 2e^- \rightarrow H_2O$ **(1)**
(iii) $C_2H_5OH + 3H_2O \rightarrow 2CO_2 + 12H^+ + 12e^-$
1 mark for correct species and balancing
1 mark for cancelling $3O_2$ and $3H_2O$ on each side
(iv) $E^{\ominus} = +1.23 - (+1.01) = +0.22\,\text{V}$ **(1)**
(b) (i) $E_{\text{cell}}^{\ominus} = +1.69 - (-0.36) = 2.05\,\text{V}$ **(1)**
(ii) $2PbSO_4 + 2H_2O \rightarrow PbO_2 + 2HSO_4^- + 2H^+ + Pb$
1 mark for correct species on each side of the equation
1 mark for balancing and cancelling H^+ and e^-
(iii) Answer above $(6 \times 2.05) = 12.30\,\text{V}$ **(1)**
There are six cells and the reactions must be reversed **(1)**.

101. Redox titrations

1 A **(1)**
2 (a) $MnO_4^- + 8H^+ + 5Fe^{2+} \rightarrow Mn^{2+} + 4H_2O + 5Fe^{3+}$ **(1)**
(b) amount of $MnO_4^- = 5.00 \times 10^{-3} \times 27.60 \times 10^{-3}$
$= 1.38 \times 10^{-4}\,\text{mol}$ **(1)**
(c) amount of Fe^{2+} in $25.0\,\text{cm}^3$ portions
$= 1.38 \times 10^{-4} \times 5 = 6.90 \times 10^{-4}\,\text{mol}$ **(1)**
(d) amount of Fe^{2+} in sample $= 10 \times 6.90 \times 10^{-4}$
$= 6.90 \times 10^{-3}\,\text{mol}$ **(1)**
(e) mass of Fe in sample $= 55.8 \times 6.90 \times 10^{-3}$
$= 0.385\,\text{g}$ **(1)**
(f) % by mass of Fe in lawnsand $= 100 \times 0.385 / 9.35$
$= 4.12\,\%$ **(1)** *Answer must be to three significant figures*
(g) It provides hydrogen ions for the redox reaction / it helps to prevent oxidation of Fe^{2+} to Fe^{3+} in air / prevents formation of brown manganese(IV) oxide **(1)**.
(h) Hydrochloric acid reacts with / is oxidised by manganate(VII) **(1)** to form chlorine gas **(1)**.

102. d-block atoms and ions

1 (a) $(1s^2\,2s^2\,2p^6\,3s^2\,3p^6)\,3d^1\,4s^2$ **(1)**
(b) $1s^2\,2s^2\,2p^6\,3s^2\,3p^6\,3d^5\,4s^1$ **(1)**
2 (a) $1s^2\,2s^2\,2p^6\,3s^2\,3p^6\,3d^5$ **(1)**
(b) $1s^2\,2s^2\,2p^6\,3s^2\,3p^6\,3d^3$ **(1)**
3 (a) d-block element **(1)** that forms one or more stable ions with incompletely-filled d-orbitals **(1)**
(b) Scandium only forms Sc^{3+} ions **(1)** which has no electrons in d-orbitals **(1)**.
4 D **(1)**

5 (a) *1 mark for correctly filled boxes, e.g.*

(b) *1 mark for correctly filled boxes, e.g.*
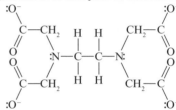

103. Ligands and complex ions

1 B **(1)**
2 (a) (i) The metal ion is Co^{2+} **(1)** the ligand is Cl^- ion **(1)**.
(ii) a species that can donate one or more pairs of electrons to a metal ion **(1)** forming dative bonds **(1)**
(b) 4 **(1)** because there are four dative bonds in the complex **(1)** *not four chloride ions*
3 It can donate one lone pair of electrons **(1)**.
not it only has one lone pair of electrons
4 *1 mark for correct structure, 1 mark for correct locations of two lone pairs of electrons, e.g.*

5 (a) Six **(1)**
(b) Locations of lone pairs of electrons correctly shown, e.g.
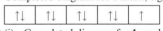
Lone pairs can be in any position next to the correct atoms.

104. Shapes of complexes

1 C **(1)**
2 (a) Square planar shape on both diagrams **(1)**
Positions of ligands correct on *cis* **(1)**   Positions of ligands correct on *trans* **(1)**
(b) square planar **(1)**
(c) The *cis* form has anti-cancer properties but the *trans* form is less effective / more toxic. **(1)**
Using the *cis* alone reduces the risk of harmful side effects. **(1)**
3 (a) octahedral **(1)** 90° **(1)**
(b) (i) tetrahedral **(1)** 109.5° **(1)**
(ii) It changes from 6 to 4. **(1)**
(iii) Chloride ions are large **(1)** compared to water molecules / aqua ligands **(1)**.

105. Colours

1 (a) Completed diagram for **1** mark, e.g.
(b) (i) Completed diagram for **1** mark, e.g.
(ii) Completed diagram for **1** mark, e.g.
(iii) Light energy is absorbed **(1)** which promotes a d electron from the lower level to the higher level. **(1)**
2 D **(1)**
3 (a) The complex contains Sc^{3+} ions **(1)** which have empty d-orbitals **(1)** so d−d transitions cannot occur **(1)**.
(b) Al^{3+} ions do not have d electrons **(1)**.

106. Colour changes

1 A **(1)**

2 (a) Multidentate **(1)**

(b) The nitrogen atoms have lone pairs of electrons **(1)** which form dative covalent / coordinate bonds with the Fe^{2+} ion **(1)**.

(c) (i) ligand exchange **(1)**

(ii) The d-orbitals in the Fe^{2+} are split **(1)**. The change in ligand changes the difference in energy between the two split levels **(1)**. A different frequency of light will be absorbed (to excite d electrons) **(1)**.

3 (a) $[Fe(H_2O)_6]^{3+}$ **(1)** there is a change in oxidation number of iron in the complex **(1)**

(b) $[Cu(NH_3)_4(H_2O)_2]^{2+}$ **(1)** there is a change in ligand in the complex **(1)**

(c) $[CoCl_4]^{2-}$ **(1)** there is a change in ligand **(1)** and coordination number **(1)**

107. Vanadium chemistry

1 (a) **C (1)**

(b) V^{2+} +2 *and* V^{3+} +3 **(1)**
VO_2^+ +5 **(1)** VO^{2+} +4 **(1)**

2 (a) (i) $2V^{3+} + Zn \rightarrow 2V^{2+} + Zn^{2+}$ **(1)**

(ii) $E_{cell}^{\ominus} = E_{right}^{\ominus} - E_{left}^{\ominus} = -0.26 - (-0.76) = +0.50\,V$ **(1)**
The value is positive so the reaction is feasible **(1)**.

(b) (i) $VO_2^+ + 4H^+ + 3e^- \rightarrow V^{2+} + 2H_2O$ **(1)** *Multiply through by 2.*
$Zn \rightarrow Zn^{2+} + 2e^-$ *Multiply through by 3, and combine.*
$2VO_2^+ + 8H^+ + 3Zn \rightarrow 2V^{2+} + 4H_2O + 3Zn^{2+}$ **(1)**

(ii) All $E_{cell}^{\ominus}$ values are positive. **(1)**

108. Chromium chemistry

1 (a) **D (1)**

(b) Cr^{2+} +2 *and* Cr^{3+} +3 **(1)**
CrO_4^{2-} +6 **(1)** $Cr_2O_7^{2-}$ +6 **(1)**

2 (a) $Cr_2O_7^{2-} + 14H^+ + 3Zn \rightarrow 2Cr^{3+} + 7H_2O + 3Zn^{2+}$ **(1)**

(b) $E_{cell}^{\ominus} = E_{right}^{\ominus} - E_{left}^{\ominus}$
$= +1.33 - (-0.76) = +2.09\,V$ **(1)**
The value is positive so the reaction is feasible **(1)**.

3 (a) $Cr^{2+} \rightarrow Cr^{3+} + e^-$ *(multiply through by 2)*
and $\frac{1}{2}O_2 + 2H^+ + 2e^- \rightarrow H_2O$ **(1)**
$2Cr^{2+} + \frac{1}{2}O_2 + 2H^+ \rightarrow 2Cr^{3+} + H_2O$ **(1)**

(b) $E_{cell}^{\ominus} = E_{right}^{\ominus} - E_{left}^{\ominus} = +1.23 - (-0.41) = +1.64\,V$ **(1)**
The value is positive so the reaction is feasible **(1)**.

4 $Cr_2O_7^{2-}(aq) + H_2O(l) \rightarrow 2CrO_4^{2-}(aq) + 2H^+(aq)$ **(1)**

109. Reactions with hydroxide ions

1 (a) (When two different solutions are mixed,) the precipitate is a solid / an insoluble product is formed **(1)**.

(b) $[Cu(H_2O)_6]^{2+}(aq) + 2OH^-(aq) \rightarrow [Cu(H_2O)_4(OH)_2](s) + 2H_2O(l)$ **1** mark for correct equation, **1** mark for correct state symbols

(c) (i) An O–H bond breaks **(1)** in a water / aqua ligand **(1)**.

(ii) proton acceptor / Brønsted–Lowry base **(1)**

2 **C (1)**

3 **1** mark for each correct colour, e.g.

$[Cu(H_2O)_6]^{2+}$	(pale) blue
$[Co(H_2O)_6]^{2+}$	(pale) pink
$[Cu(H_2O)_4(OH)_2]$	blue
$[Co(H_2O)_4(OH)_2]$	blue

4 (a) green **(1)** brown **(1)**   (b) oxidation / redox **(1)**

110. Reactions with ammonia

1 (a) (i) $[Cu(H_2O)_6]^{2+}(aq) + 2NH_3(aq)$
$\rightarrow [Cu(H_2O)_4(OH)_2](s) + 2NH_4^+(aq)$
1 mark for correct equation,
1 mark for correct state symbols

(ii) acid–base reaction **(1)**

(iii) Brønsted–Lowry base / proton acceptor **(1)** it accepts a proton from water molecules / aqua ligands **(1)** when O–H bonds break **(1)**.

(b) (i) $[Cu(NH_3)_4(H_2O)_2]^{2+}$ **(1)** *not $[Cu(NH_3)_6]^{2+}$*

(ii) ligand **(1)**

2 A **(1)**
precipitate formed by cobalt(II) hydroxide re-dissolves but the one formed by iron(II) hydroxide does not

3 (a) **1** mark for each correct colour, e.g.

$[Co(H_2O)_4(OH)_2]$	blue
$[Cr(H_2O)_3(OH)_3]$	grey-green
$[Co(NH_3)_6]^{2+}$	brown
$[Cr(NH_3)_6]^{3+}$	purple

(b) (i) $[Co(H_2O)_4(OH)_2](s) + 6NH_3(aq) \rightarrow$
$[Co(NH_3)_6]^{2+}(aq) + 4H_2O(l) + 2OH^-(aq)$
1 mark for correct equation, **1** mark for correct state symbols

(ii) $[Cr(H_2O)_3(OH)_3](s) + 6NH_3(aq)$
$\rightarrow [Cr(NH_3)_6]^{3+}(aq) + 3H_2O(l) + 3OH^-(aq)$
1 mark for correct equation, **1** mark for correct state symbols

111. Ligand exchange

1 (a) (i) (large) increase **(1)**

(ii) There are more particles on the right / 2 on the left and 7 on the right **(1)**.
The system becomes more disordered **(1)**.

(iii) To reverse the reaction would need a decrease in ΔS_{system} **(1)** which is less likely to happen **(1)**.

(b) There are equal numbers of (similar size and complexity) particles on each side of the equation **(1)** so there is little change in ΔS_{system} **(1)**.

2 (a) Correctly completed table: **1** mark for each shape, **1** mark for each bond angle, e.g.

Complex ion	$[Co(NH_3)_6]^{2+}$	$[CoCl_4]^{2-}$
Shape		
Name of shape	octahedral **(1)**	tetrahedral **(1)**
Coordination number	6 **(1)**	4 **(1)**

(b) They have different ligands **(1)** and different coordination numbers **(1)** so the d–d splitting is different / different splitting of energy levels **(1)**.

112. Heterogeneous catalysis

1 (A heterogeneous catalyst is in a) different phase / state from the reactants **(1)**.

2 **C (1)**

3 (a) (i) $V_2O_5 + SO_2 \rightarrow V_2O_4 + SO_3$ **(1)**

(ii) $V_2O_4 + \frac{1}{2}O_2 \rightarrow V_2O_5$ **(1)**

(b) The oxidation number of vanadium decreases from +5 to +4 in the first reaction **(1)** and increases to +5 again in the second reaction **(1)**, overall V_2O_5 is unchanged (but SO_2 is oxidised to SO_3) **(1)**.

4 (a) Carbon monoxide is toxic **(1)** and nitrogen monoxide is a cause of acid rain / it is a NO_x / causes breathing difficulties **(1)**.

(b) $CO + NO \rightarrow CO_2 + \frac{1}{2}N_2$ / $2CO + 2NO \rightarrow 2CO_2 + N_2$ **(1)**

(c) $C_8H_{18} + 12\frac{1}{2}O_2 \rightarrow 8CO_2 + 9H_2O$ /
$2C_8H_{18} + 25O_2 \rightarrow 16CO_2 + 18H_2O$ **(1)**

113. Homogeneous catalysis

1 (A homogeneous catalyst is in the) same phase / state as the reactants **(1)**.

2 (a) $S_2O_8^{2-} + 2I^- \rightarrow 2SO_4^{2-} + I_2$ **(1)**

(b) The ions are both negatively charged / have the same charge (so they repel each other). **(1)**

(c) (i) $2Fe^{3+} + 2I^- \rightarrow 2Fe^{2+} + I_2$ **(1)**

(ii) $2Fe^{2+} + S_2O_8^{2-} \rightarrow 2Fe^{3+} + 2SO_4^{2-}$ **(1)**

(d) The oxidation number of iron decreases from +3 to +2 in the first reaction **(1)** and increases to +3 again in the second reaction **(1)**, overall iron(III) is unchanged **(1)**, ions are oppositely charged (so lower activation energy **(1)**.

(e) Iron(III) is converted to iron(II) and back to iron(III) again **(1)**, so iron(II) will also catalyse the reaction **(1)**.

3 (a) The reaction is catalysed by one of its products **(1)**.
  (b) Mn^{2+} **(1)**

114. Exam skills 10

1 (a) (i) $1s^2\ 2s^2\ 2p^6\ 3s^2\ 3p^6\ 3d^6\ 4s^2$ **(1)**
   (ii) $1s^2\ 2s^2\ 2p^6\ 3s^2\ 3p^6\ 3d^6$ **(1)**
  (b) Its d-orbitals are split in energy **(1)** by ligands **(1)**, visible light is absorbed to allow electron transitions to a higher energy level **(1)**.
  (c) Scandium atoms have an incompletely-filled d-orbital **(1)** but do not form stable ions with incompletely filled d-orbitals / Sc^{3+} ions have no occupied d-orbitals **(1)**

2 (a) $[Cu(H_2O)_6]^{2+} + EDTA^{4-} \rightarrow [Cu(EDTA)]^{2-} + 6H_2O$ **(1)** ligand exchange **(1)**
  (b) The forward reaction involves an increase in ΔS_{system} **(1)**, which is more likely to happen than a decrease in the backward reaction **(1)**.
  (c) $EDTA^{4-}$ can donate several / six lone pairs of electrons **(1)** to form dative bonds with the metal ion **(1)**.

3 (a) $[Cr(H_2O)_3(OH)_3](s) + 3OH^-(aq) \rightleftharpoons [Cr(OH)_6]^{3-}(aq) + 3H_2O(l)$ **(1)**
   $[Cr(H_2O)_3(OH)_3](s) + 3H_3O^+(aq) \rightleftharpoons [Cr(H_2O)_6]^{3+}(aq) + 3H_2O(l)$ **(1)**
  (b) Chromium(III) hydroxide reacts with bases / OH^- ions **(1)** and with acids / H^+ ions /H_3O^+ ions **(1)**.

115. Measuring reaction rates

1 (The rate of a reaction is the change in) the concentration of a product or a reactant per unit of time. **(1)**

2 (a) Hydrogen: $2.0 \times 5.0 \times 10^{-3} = 0.010\,g$ **(1)**
   Carbon dioxide: $44.0 \times 5.0 \times 10^{-3} = 0.22\,g$ **(1)**
  (b) $24 \times 1000 \times 5.0 \times 10^{-3} = 120\,cm^3$ **(1)**
  (c) (i) A small amount of gas can be measured on the balance **(1)** a small amount occupies a large volume **(1)**.
   (ii) Use a gas syringe **(1)** a large amount of gas would need to be produced for it to be measured on the balance / a small amount of gas occupies a large volume **(1)**.

3 C **(1)**

4 (a) to slow the reaction down / quench the reaction **(1)**
  (b) (Draw the tangent and) calculate the gradient **(1)** at the start **(1)**.

116. Rate equation and initial rate

1 (The order with respect to a substance in a rate equation is the) power to which that reactant's concentration is raised. **(1)**

2 (a) *Correctly completed table, 1 mark for each correct box*

Experiment	Initial [X] / mol dm⁻³	Initial [Y] / mol dm⁻³	Initial rate / mol dm⁻³ s⁻¹
1	*0.10*	*0.10*	*5.0 × 10⁻³*
2	*0.40*	*0.10* **(1)**	*2.0 × 10⁻²*
3	*0.10*	*0.40*	*8.0 × 10⁻²* **(1)**
4	*0.02*	*0.30* **(1)**	*9.0 × 10⁻³*

  (b) (i) $k = \text{rate}/([X][Y]^2)$
    $k = (5.0 \times 10^{-3}) / (0.10 \times 0.1^2)$ **(1)**
    $k = 5.0$ **(1)**
   (ii) $mol^{-2}\ dm^6\ s^{-1}$ **(1)**
  (c) change in temperature **(1)**

3 (a) A: first order **(1)**, B: second order **(1)**, C: zero order **(1)**
  (b) rate $= k[A][B]^2$ / rate $= k[A][B]^2[C]^0$ **(1)**
  (c) third order **(1)**

117. Rate equation and half-life

1 1 mark for each correct line, e.g.

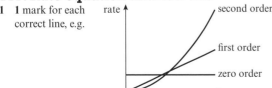

2 (a) 1 mark for each of the following points:
  • sensible scales for both axes (i.e. plotted points will cover > 50%)
  • both axes labelled with quantity and unit
  • all points plotted ±1 mm
  • single, smooth line of best fit

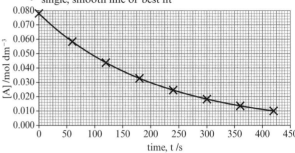

  (b) (i) $(200 - 55) = 145\,s$ **(1)**   (ii) $(345 - 200) = 145\,s$ **(1)**
  (c) First order with respect to reactant A **(1)** because the half-lives are the same / half-life is constant **(1)**.
  (d) (Tangent to curve drawn at 100 s)
   Values for $\Delta[A]$ and Δt determined and used to calculate instantaneous rate
   For example:
   $\Delta[A] = (0.062 - 0.040) = 0.022\,mol\,dm^{-3}$ **(1)**
   $\Delta t = (40 - 220) = -180\,s$ **(1)**
   Instantaneous rate $= \dfrac{-(0.022)}{(-180)} = 1.22 \times 10^{-4}\,mol\,dm^{-3}\,s^{-1}$ **(1)**

118. Rate-determining steps

1 The slowest step of a chemical reaction **(1)**.

2 (a) $(CH_3)_3CBr + OH^- \rightarrow (CH_3)_3COH + Br^-$ **(1)**
  (b) (i) In Step 1 a covalent bond must be broken **(1)**, but in Step 2 two oppositely charged ions attract each other (and form a dative covalent bond) **(1)**.
   (ii) The overall rate does not depend on the concentration of hydroxide ions / the rate-determining step is not one that involves hydroxide ions **(1)** so the rate is zero order with respect to hydroxide ions **(1)**.
  (c) S_N1 mechanism **(1)** because only one reactant is involved in the rate-determining step **(1)**

3 Step 2 **(1)** because X and Y are in the same amounts in that step as they are in the rate equation **(1)**.

119. Finding the activation energy

1 (a) Correctly completed table, **1** mark for each correct column, e.g.

Temperature, T /K	Rate constant, k /mol dm⁻³ s⁻¹	1/T /K⁻¹	ln k
656	*1.39 × 10⁻⁴*	*1.52 × 10⁻³*	*−8.88*
695	*9.52 × 10⁻⁴*	*1.44 × 10⁻³*	*−6.96*
735	*5.53 × 10⁻³*	*1.36 × 10⁻³*	*−5.20*
788	*4.32 × 10⁻²*	*1.27 × 10⁻³*	*−3.14*
846	*3.05 × 10⁻¹*	*1.18 × 10⁻³*	*−1.19*

  (b) **1** mark for each of the following points:
  • sensible scales for both axes (i.e. plotted points will cover > 50%)
  • both axes labelled with quantity and unit
  • all points plotted ±1 mm
  • single, straight line of best fit

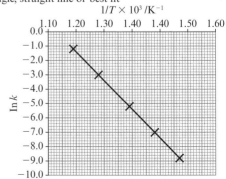

(c) Correct values for change in $\ln k$ and $1/T$ for **1** mark, e.g.
change in $\ln k = -8.88 - (-1.19) = -7.69$
change in $1/T = (1.52 - 1.18) \times 10^3 = 0.34 \times 10^{-3}$
Calculation of gradient for **1** mark, e.g.
$(-7.69)/(0.34 \times 10^{-3}) = -22\,618$

(d) $-E_a/R = -22\,618$
$E_a = 22\,618 \times 8.31 / 1000$ **(1)**
$= +188\,kJ\,mol^{-1}$ **(1)**

120. Exam skills 11

1 (a) **1** mark for each of the following points:
- sensible scales for both axes (i.e. plotted points will cover > 50%)
- both axes labelled with quantity and unit
- all points plotted ±1 mm
- single, curved line of best fit

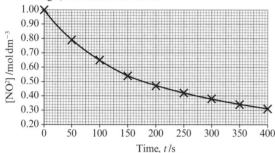

(b) If the reaction was zero order the rate would not be affected by $[NO_2]$ **(1)**, so the gradient would be constant **(1)**, if the gradient changes with $[NO_2]$ the reaction must be first or second order **(1)**.

(c) (i) Tangent to curve drawn at $[NO_2] = 0.70\,mol\,dm^{-3}$ **(1)**
Gradient calculated (**1** mark for correct values from the graph, **1** mark for calculation)
For example, $(0.30 - 0.92) / (240 - 0)$ **(1)**
$= -2.6 \times 10^{-3}\,mol\,dm^{-3}\,s^{-1}$ **(1)**

(ii) Determine the gradient / instantaneous rate of reaction at different values for $[NO_2]$. **(1)**
Plot a graph of gradient / instantaneous rate of reaction against the square of $[NO_2]$. **(1)**
If the gradient of this graph is constant and positive, the order is second order. **(1)**

121. Identifying aldehydes and ketones

1 (a) **1** mark for each correctly completed box, e.g.

Name of compound	3-methylbutanal	pentan-2-one
Warmed with acidified potassium dichromate(VI) solution	orange solution to green solution	no change / solution stays orange
Warmed with Fehling's solution	blue solution to red precipitate	no change / solution stays blue
2,4-dinitrophenyl-hydrazine	orange-yellow precipitate	orange-yellow precipitate

(b) no differences **(1)**

2 **B (1)**

3 (a) Boiling temperature increases as number of electrons increases / the molar mass increases **(1)** because the London forces (between molecules) increase **(1)**.

(b) Their molecules form hydrogen bonds **(1)** with water molecules **(1)**.

122. Optical isomerism

1 **B (1)**

2 (a) (i) 1-chlorobutane **(1)**
Displayed formula **(1)**, e.g.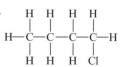

(ii) 2-chlorobutane **(1)**
Displayed formula **(1)**, e.g.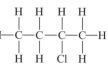

(b) It contains a chiral centre **(1)** asymmetric carbon atom / 4 different groups attached **(1)**.

(c) equal amounts / moles **(1)** of each optical isomer / enantiomer **(1)**

(d) Vibrations / oscillations (of light waves) in one direction / one plane only **(1)**.
Light is of a single wavelength / frequency **(1)**.

3 Rotate analyser to maximum darkness **(1)**, place solution of compound between polariser and analyser **(1)**, rotate analyser to obtain maximum darkness **(1)**, measure angle of rotation **(1)**.

123. Optical isomerism and reaction mechanisms

1 **D (1)**

2 (a) (trigonal) planar **(1)** 120° **(1)**

(b) It can approach from both sides / above and below. **(1)**

(c) The optical isomer formed depends upon how the cyanide ion approached the electron-deficient / partially positively charged carbon atom **(1)**, there are is an equal chance of this being on one side or the other **(1)**.

3 **B (1)**

124. Reactions of aldehydes and ketones

1 (a) **D (1)**

(b) **1** mark for each correctly completed box to a maximum of three marks.
*Deduct **1** mark for each answer more than three answers.*

Compound	Forms precipitate?
CH_3OH	
CH_3CH_2OH	✓
CH_3CHO	✓
CH_3CH_2CHO	
$CH_3COCH_2CH_2CH_3$	✓
$CH_3CH_2COCH_2CH_3$	

2 (a) ether **(1)** / <u>dry</u> (ether) **(1)**

(b) (i) $CH_3CH_2COCH_3 + 2[H] \rightarrow CH_3CH_2CH(OH)CH_3$ **(1)**
butan-2-ol **(1)**

(ii) $CH_3CH_2CH_2CHO + 2[H] \rightarrow CH_3CH_2CH_2CH_2OH$ **(1)**
butan-1-ol **(1)**

3 (a) nucleophilic addition **(1)**
Both words needed for the mark

(b) *Reaction mechanism, e.g.*

1 mark for two curly arrows on left-hand diagram
1 mark for correct intermediate in centre diagram
1 mark for two curly arrows in centre diagram

125. Carboxylic acids

1 **B (1)**

2 (a) (i) sulfuric acid **(1)**
   (ii) ethanal **(1)**
   (iii) $CH_3CH_2OH + 2[O] \rightarrow CH_3COOH + H_2O$
   All formulae correct **(1)** balanced **(1)**
   (b) (i) $CH_3CN + H^+ + 2H_2O \rightarrow CH_3COOH + NH_4^+$ /
   $CH_3CN + HCl + 2H_2O \rightarrow CH_3COOH + NH_4Cl$
   all formulae correct **(1)** balanced **(1)**
   (ii) reflux **(1)**

3 $2CH_3CH_2COOH + CaCO_3 \rightarrow (CH_3CH_2COO)_2Ca + H_2O + CO_2$ **(1)**
   calcium propanoate **(1)**

4 (a) $CH_3COOH + PCl_5 \rightarrow CH_3COCl + POCl_3 + HCl$
   correct organic product **(1)** other correct species and balanced **(1)**
   (b) distillation **(1)**

126. Making esters

1 **C (1)**

2 (a) (i) ethanol **(1)** methanoic acid **(1)**
   (ii) $CH_3CH_2OH + HCOOH \rightarrow HCOOCH_2CH_3 + H_2O$ **(1)**
   (iii) methanoyl chloride **(1)**
   (iv) hydrogen chloride **(1)**
   (b) (i) methyl ethanoate **(1)**
   Displayed formula for **1** mark, e.g.

   $$\begin{array}{ccccc} & H & & O & H \\ & | & & \| & | \\ H - & C & - O - & C & - C - H \\ & | & & & | \\ & H & & & H \end{array}$$

   (ii) propanoic acid **(1)**
   Displayed formula for **1** mark, e.g.

   $$\begin{array}{ccc} H & H & O \\ | & | & \\ H - C - C - C & \\ | & | & \\ H & H & O-H \end{array}$$

3 (a) (i) breaking down a molecule by reaction with water **(1)**
   (ii) $CH_3CH_2COOCH_2CH_3 + H_2O \rightleftharpoons CH_3CH_2COOH + CH_3CH_2OH$
   1 mark for correct species and balanced, **1** mark for reversible symbol
   (b) $CH_3CH_2COOCH_2CH_3 + NaOH \rightarrow CH_3CH_2COONa + CH_3CH_2OH$ **(1)**

127. Making polyesters

1 Similarity for **1** mark, e.g. both form a long chain of repeating units / both involve reactions between monomers
   Difference for **1** mark, e.g. in addition polymerisation the polymer is the only product but in condensation polymerisation a smaller molecule / water / HCl is also produced

2 **D (1)**

3 (a) butane-1,4-diol **(1)** and butanedioic acid **(1)**
   (b) Correct repeat unit, e.g.

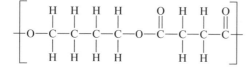

   Ester link including C=O **(1)**
   Rest of repeat unit with O at correct end **(1)**

128. Benzene

1 (a) C_6H_6 **(1)**
   (b) (i) a **smoky** flame **(1)**
   (ii) $C_6H_6 + 4\frac{1}{2}O_2 \rightarrow 2C + 2CO + 2CO_2 + 3H_2O$
   1 mark for all correct species, **1** mark for correct balancing
   (c) aluminium bromide / iron(III) bromide **(1)**

2 (a) Two p-orbitals overlap **(1)** sideways **(1)** once a σ / sigma bond has formed **(1)**.
   (b) three **(1)**

3 (a) $3 \times (-120) = -360\,kJ\,mol^{-1}$ **(1)**
   (b) (i) more stable by $152\,kJ\,mol^{-1}$ **(1)**
   (ii) p electrons / π bonds are delocalised **(1)**

129. Halogenation of benzene

1 (a) **B (1)**
   (b) Benzene has delocalised π bonds **(1)**, alkenes have localised electron density in their π bond **(1)**.

2 (a) (i) Br^+ **(1)**
   (ii) $AlBr_3 + Br_2 \rightarrow [AlBr_4]^- + Br^+$ **(1)**
   (iii) $[AlBr_4]^- + H^+ \rightarrow AlBr_3 + HBr$ **(1)**
   (b) Reaction mechanism, e.g.

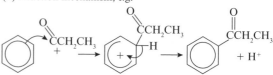

   1 mark for arrow from circle to Br⁺
   1 mark for intermediate (part circle no further than limits shown)
   1 mark for arrow from C−Br bond

3 lone pair electrons on the oxygen atom (in the hydroxyl group) **(1)** delocalises with electrons in the delocalised π bond **(1)** increases electron density of benzene ring **(1)**

130. Nitration of benzene

1 (a) **concentrated** sulfuric acid **(1) concentrated** nitric acid **(1)**
   (b) (i) $^+NO_2$ / NO_2^+ **(1)**
   (ii) $HNO_3 + H_2SO_4 \rightarrow NO_2^+ + HSO_4^- + H_2O$ **(1)**
   (iii) Reaction mechanism, e.g.

   1 mark for arrow from circle to +NO₂
   1 mark for intermediate (part circle no further than limits shown)
   1 mark for arrow from C−H bond

2 (a) Correct structure for **1** mark, e.g.

   (b) Structures can be shown in either order, e.g.

Structure		
	NO₂ NO₂ structure **(1)**	NO₂ NO₂ structure **(1)**
Name	1,2-dinitrobenzene **(1)**	1,4-dinitrobenzene **(1)**

   (c) The nitro group contains electronegative elements / nitrogen / oxygen **(1)** which withdraw electron density from the delocalised π bond / deactivate the ring **(1)**.

131. Friedel−Crafts reactions

1 (a) $CH_3CH_2COCl + AlCl_3 \rightarrow CH_3CH_2C^+O + [AlCl_4]^-$ **(1)**
   $[AlCl_4]^- + H^+ \rightarrow AlCl_3 + HCl$ **(1)**
   (b) Reaction mechanism, e.g.

   1 mark for arrow from circle to + on electrophile
   1 mark for intermediate (part circle no further than limits shown)
   1 mark for arrow from C−H bond

(c) (i) The aluminium atom has a vacant orbital. **(1)**

 (ii) NH_4Cl could not act as a catalyst for this reaction **(1)**, there is no vacant orbital (on N) / it cannot accept a pair of electrons **(1)**.

2 (a) anhydrous **(1)** reflux **(1)**

 (b) Reaction mechanism, e.g.

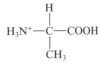

 1 mark for arrow from circle to + on electrophile
 1 mark for intermediate (part circle no further than limits shown)
 1 mark for arrow from C−H bond

132. Making amines

1 **B (1)**

2 From left to right, **1** mark for each correct answer: primary, secondary, tertiary, primary

3 (a) lithium tetrahydridoaluminate / lithium aluminium hydride **(1)**

 (b) $CH_3CN + 4[H] \rightarrow CH_3CH_2NH_2$ **(1)**

4 (a) reflux **(1)**

 (b) $C_6H_5NO_2 + 6[H] \rightarrow C_6H_5NH_2 + 2H_2O$
 1 mark for correct species, **1** mark for balancing

5 (a) $CH_3CH_2OH + 2NH_3 \rightarrow CH_3CH_2NH_2 + NH_4Br$ **(1)**

 (b) The (organic) product formed acts as a nucleophile **(1)**, so further substitution can happen **(1)**, excess ammonia reduces the chance of further substitution / production of secondary amines **(1)**.

133. Amines as bases

1 **D (1)**

2 (a) proton acceptor **(1)**

 (b) $NH_3(aq) + H_2O(l) \rightleftharpoons NH_4^+(aq) + OH^-(aq)$ **(1)**
 Ammonia molecule accepts a proton / hydrogen ion from water molecule. **(1)**

3 (a) lone pair of electrons on nitrogen atom **(1)**

 (b) Primary aliphatic amines are stronger bases than ammonia **(1)**.
 Amines have alkyl groups which are electron-releasing **(1)** and increase electron density on nitrogen atom / make lone pair of electrons more available (to form a bond with hydrogen ion) **(1)**.

 (c) Aromatic amines are weaker bases than ammonia **(1)**.
 Lone pair of electrons on nitrogen atom delocalises with π electrons in the benzene ring **(1)**, which decreases electron density on nitrogen atom / make lone pair of electrons less available (to form a bond with hydrogen ion) **(1)**.

4 (a) Amine molecules can form hydrogen bonds with water molecules. **(1)**

 (b) $C_3H_7NH_2 + H_2O \rightarrow C_3H_7NH_3^+ + OH^-$ **(1)**

 (c) $C_3H_7NH_2 + HCl \rightarrow C_3H_7NH_3Cl$ **(1)** propylammonium chloride **(1)**

134. Other reactions of amines

1 **A (1)**

2 (a) (i) $2C_3H_7NH_2(aq) + [Cu(H_2O)_6]^{2+}(aq) \rightarrow$
  $2C_3H_7NH_3^+(aq) + [Cu(H_2O)_4(OH)_2](s)$
  1 mark for correct species, **1** mark for balancing, **1** mark for state symbols

  (ii) Acid−base **(1)**

 (b) (i) $4C_3H_7NH_2(aq) + [Cu(H_2O)_4(OH)_2](s) \rightarrow$
  $[Cu(C_3H_7NH_2)_4(H_2O)_2]^{2+}(aq) + 2H_2O(l) + 2OH^-(aq)$
  1 mark for correct species, **1** mark for balancing, **1** mark for state symbols

  (ii) ligand exchange **(1)**

3 (a) (i) chloroethane **(1)** CH_3CH_2Cl **(1)**

  (ii) $CH_3NH_2 + CH_3CH_2Cl \rightarrow CH_3NHCH_2CH_3$ **(1)**

 (b) secondary **(1)**

 (c) There is a lone pair of electrons on the nitrogen atom of *N*-methylethylamine **(1)** so *N*-methylethylamine acts as a nucleophile **(1)** and further substitution happens with chloroethane **(1)**.

135. Making amides

1 **A (1)**

2 (a) NH_2 group circled and labelled A **(1)**

 (b) $CONH_2$ group circled and labelled B **(1)**

3 (a) CH_3CH_2COCl **(1)** propanoyl chloride **(1)**

 (b) (i) $CH_3CH_2COCl + 2NH_3 \rightarrow CH_3CH_2CONH_2 + NH_4Cl$ **(1)**

  (ii) $CH_3CH_2COCl + CH_3CH_2NH_2 \rightarrow$
  $CH_3CH_2CONHCH_2CH_3 + HCl$ / $CH_3CH_2COCl +$
  $2CH_3CH_2NH_2 \rightarrow CH_3CH_2CONHCH_2CH_3 +$
  $CH_3CH_2NH_3Cl$ **(1)**

 (c) propanamide **(1)**

4 (a) amide because it is a derivative of ammonia where hydrogen atoms are replaced by alkyl groups **(1)**, secondary because two hydrogen atoms are replaced by alkyl groups **(1)**

 (b) propyl because an alkyl group with three carbon atoms **(1)**, *N* as joined to a nitrogen atom **(1)**, ethanamide because amide functional group is part of a group with two carbon atoms **(1)**

136. Making polyamides

1 **C (1)**

2 (a) butane-1,4-diamine **(1)**

 (b) faster reaction **(1)**

 (c) Repeat unit, e.g.

 Amide link including C=O and N−H **(1)**
 Rest of repeat unit with N and O at correct ends **(1)**

137. Amino acids

1 (a) 2-amino-3-methylbutanoic acid **(1)**

 (b) It contains an amino group / −NH_2 group **(1)** and a carboxyl group /−COOH **(1)**.

 (c) Valine has optical isomers **(1)** because carbon atom 2 **(1)** has four different groups attached / is asymmetric / has a chiral centre **(1)**.

2 **D (1)**

3 (a) (i) *Correct structure* (ii) *Correct structure*
   for 1 mark   *for 1 mark*

 (b) neutral molecule **(1)** with $^+NH_3$ group and COO^- group **(1)**

 (c) $NH_2CH(CH_3)COOH + KOH \rightarrow NH_2CH(CH_3)COO^-K^+ + H_2O$ **(1)**

138. Proteins

1 **B (1)**

2 (a) Displayed formula **(1)**

 (b) Box drawn around CONH group **(1)**

(c) (i) Repeat unit, e.g.

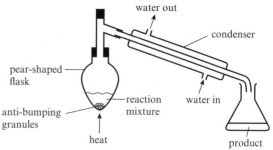

Peptide link including C=O and N–H (**1**)
Rest of repeat unit with N and O at correct ends (**1**)
(ii) condensation (**1**)
3 (a) concentrated / 6 mol dm^{-3} (**1**) hydrochloric acid (**1**) heat / reflux (**1**)
(b) H$_2$NCH(CH$_2$OH)CONHCH(CH$_2$OH)COOH + 2H$^+$ + H$_2$O $\rightarrow$ 2H$_3\overset{+}{N}$CH(CH$_2$OH)COOH
1 mark for correct species, 1 mark for balancing

139. Exam skills 12
1 (a) Step 1: CH$_3$CH$_2$CH$_2$Cl + KCN $\rightarrow$ CH$_3$CH$_2$CH$_2$CN + KCl (**1**)
Step 2: CH$_3$CH$_2$CH$_2$CN + 2H$_2$ $\rightarrow$ CH$_3$(CH$_2$)$_3$NH$_2$ / CH$_3$CH$_2$CH$_2$CN + 4[H] $\rightarrow$ CH$_3$(CH$_2$)$_3$NH$_2$ (**1**)
(b) butanenitrile (**1**)
2 (a) There is a lone pair of electrons on the nitrogen atom (**1**) which can form a (dative) bond with a hydrogen ion / proton (**1**).
(b) Ammonia is the stronger base (**1**), in phenylamine the lone pair of electrons on the N atom delocalises with the delocalised π electrons in the benzene ring (**1**) which decreases the electron density / lone pair of electrons becomes less available (**1**).
(c) C$_6$H$_5$NH$_2$ + H$_2$O $\rightleftharpoons$ C$_6$H$_5$NH$_3$$^+$ + OH$^-$ (**1**)
Water is acting as a Brønsted–Lowry acid (**1**) because it donates a proton / H$^+$ ion to phenylamine (**1**).
3 (a) 2-aminopropanoic acid (**1**)
(b) **1** mark for each dipeptide, e.g.

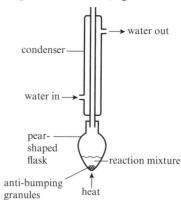

140. Grignard reagents
1 (a) magnesium (**1**)
(b) halogenoalkane (**1**)
(c) reflux (**1**) ether (**1**)
(d) to prevent the Grignard reagent reacting with water (**1**)
2 **1** mark for each correctly completed box, e.g.

Desired product	Grignard reagent with
primary alcohol	*methanal*
secondary alcohol	aldehyde (not methanal) (**1**)
tertiary alcohol	ketone (**1**)
carboxylic acid	carbon dioxide (**1**)

3 (a) CH$_3$CH$_2$CH$_2$Br + Mg $\rightarrow$ CH$_3$CH$_2$CH$_2$MgBr (**1**)
(b) CH$_3$CH$_2$CH$_2$MgBr + HCHO $\rightarrow$ CH$_3$CH$_2$CH$_2$CH$_2$OMgBr (**1**)
(c) CH$_3$CH$_2$CH$_2$CH$_2$OMgBr + H$_2$O $\rightarrow$ CH$_3$CH$_2$CH$_2$CH$_2$OH + Mg(OH)Br
1 mark for H$_2$O and CH$_3$CH$_2$CH$_2$CH$_2$OH, **1** mark for Mg(OH)Br (and balanced)
(d) product has one more C atom (**1**) different functional group / OH not Br (**1**)

141. Methods in organic chemistry 1
1 (a) It is corrosive (**1**) wear gloves (**1**) to avoid skin contact (**1**).
(b) It is quick / quicker. (**1**)
(c) (i) Crystals might not form if excess solvent is used / too much product would be in solution. (**1**)

(ii) Insoluble impurities are removed by filtration. (**1**)
(iii) Soluble impurities are left in the filtrate / stay in solution (after the aspirin has crystallised). (**1**)
(d) Measure the melting temperature of the product (**1**) product is pure if melting point is sharp (**1**).

142. Methods in organic chemistry 2
1 C (**1**)
2 (a) Diagram to show simple distillation:

- flask (pear-shaped or round-bottomed) with heat source (**1**)
- condenser attached to still head and with collection vessel (**1**)
- water flow shown correctly (**1**)
Make sure that there are no gaps, the top of the still head is closed, and the end of the condenser is open.
(b) Diagram to show reflux, e.g.

flask (pear-shaped or round-bottomed) with heat source (**1**)
- condenser vertical (**1**)
- water flow shown correctly (**1**)
Make sure that there are no gaps and the top of the condenser is open.
3 (a) Steam is passed into the reaction mixture / crushed plant material (**1**).
The product / lavender oil leaves with the water vapour (**1**) and is condensed (**1**).
(b) A lower temperature is needed / the product is less likely to decompose (**1**).

143. Reaction pathways
1 Step 1: steam (**1**) phosphoric acid catalyst (**1**) at 300 °C / 7 MPa pressure (**1**)
or sulfuric acid (**1**) then add water (**1**) and warm (**1**)
Step 2: K$_2$CrO$_7$ (**1**) acidified with dilute sulfuric acid (**1**) reflux (**1**)
2 Step 1: ethanoyl chloride (**1**) aluminium chloride (**1**) reflux (**1**)
Step 2: LiAlH$_4$ (**1**) in **dry** ether (**1**)
Step 3: phosphoric acid (**1**) heat to 180 °C (**1**) *or* Al$_2$O$_3$ (**1**) heat to 300 °C (**1**)
3 Sensible advantage and disadvantage for each route, for example:
ethene + chlorine, single product formed / 100% atom economy (**1**) but ethene must be manufactured (**1**)
ethane + chlorine, ethane obtained from crude oil (**1**) but several products are made (**1**)

144. Chromatography

1 (a) (i) 25 mm
    (ii) 16.5 mm
    Both required for the mark
  (b) $\dfrac{\text{Answer (i)}}{\text{Answer (ii)}} = 66\%$ **(1)**

2 (a) It did not **(1)** because the peak for Sudan I was not seen in the chromatogram for the stock chilli powder **(1)**.
  (b) Krakatoa chilli was contaminated **(1)** because its (it was the only one) chromatogram that showed a peak at the same position as Sudan I **(1)** mention of retention time **(1)**.
  (c) (i) There was only one peak in the chromatogram for Sudan I **(1)**
    (ii) Mass spectrometry **(1)** *this is the method indicated by the Specification*

145. Functional group analysis

1 D **(1)**

2 (a) Add bromine / bromine water **(1)**
      with hex-1-ene this changes from orange-brown to colourless **(1)**
      with hexane there is no change **(1)**
      or
      Add $KMnO_4$ acidified with dilute sulfuric acid **(1)**
      with hex-1-ene it changes from purple to colourless **(1)**
      with hexane there is no change / stays purple **(1)**
  (b) (Add $K_2Cr_2O_7$ acidified with) dilute sulfuric acid **(1)**
      with 2-methylpropan-2-ol there is no change / stays orange **(1)** with 2-methylpropan-1-ol it changes from orange to green **(1)**
  (c) Add sodium carbonate / magnesium **(1)**
      with propanoic acid there is bubbling **(1)**
      with methyl ethanoate there is no change **(1)**
      or
      Add universal indicator solution **(1)**
      with propanoic acid it changes from green to orange / yellow / red **(1)**
      with methyl ethanoate there is no change / stays green **(1)**
  (d) Add Fehling's solution / Benedict's solution **(1)**
      with propanal there is a red precipitate / turns orange **(1)**
      with propanone there is no change / solution stays blue **(1)**
      or
      Add Tollens' reagent **(1)**
      with propanal there is a silver mirror **(1)**
      with propanone there is no change / solution stays colourless **(1)**
      or
      Add $KMnO_4$ acidified with dilute sulfuric acid **(1)**
      with propanal it changes from orange to green **(1)**
      with propanone there is no change / solution stays orange **(1)**

3 Add a mixture of ethanol and sodium hydroxide solution **(1)** warm it **(1)** acidify with dilute nitric acid **(1)** add silver nitrate solution **(1)** white precipitate for 1-chloropropane **(1)** cream precipitate for 1-bromopropane **(1)** yellow precipitate for 1-iodopropane **(1)**

146. Combustion analysis

1 (a) $CH_4 + 2O_2 \rightarrow CO_2 + 2H_2O$ **(1)**
  (b) (i) $100\,cm^3$ **(1)** (ii) $50\,cm^3$ **(1)**

2 D **(1)**

3 (a) $4.39 \times 12.0 \div 44.0 = 1.20\,g$ **(1)**
      $2.25 \times 2 \times 1.0 \div 18.0 = 0.25\,g$ **(1)**
      $2.25 - (1.20 + 0.25) = 0.80\,g$ **(1)**
  (b) Calculation of empirical formula, e.g.

C	H	O	**(1)**
$0.10 \div 0.050$	$0.25 \div 0.050$	$0.050 \div 0.050$	**(1)**
$= 2$	$= 5$	$= 1$	

      Empirical formula is C_2H_5O **(1)**
  (c) molar mass of $C_2H_5O = 45\,g\,mol^{-1}$
      factor is $90/45 = 2$ **(1)**
      molecular formula is $C_4H_{10}O_2$ **(1)**

147. High-resolution mass spectra

1 D **(1)**

2 (a) (i) $(1 \times 12.0107) + (2 \times 1.0078) + (2 \times 15.9994)$
        $= 46.0251$ **(1)**
    (ii) $(2 \times 12.0107) + (6 \times 1.0078) + (1 \times 15.9994)$
        $= 46.0676$ **(1)**
  (b) The fragmentation pattern **(1)** will be different for the two compounds **(1)**.

3 (a) (i) $(3 \times 12.0107) + (7 \times 1.0078) + (1 \times 35.4532)$
        $= 78.5399$ **(1)**
    (ii) $(2 \times 12.0107) + (3 \times 1.0078) + (1 \times 15.9994)$
        $+ (1 \times 35.4532) = 78.4974$ **(1)**
  (b) Both compounds have the same number of atoms of each element / same molecular formula **(1)**.

148. ^{13}C NMR spectroscopy

1 A **(1)**

2 (a) (i) 3 **(1)** (ii) C–CHO **(1)** (iii) C–C **(1)**
  (b) 2 peaks **(1)** two identical CH_3 groups **(1)** one C=O group **(1)**

3 **1** mark for chemical environments identified, to **2** marks for both structures, e.g.

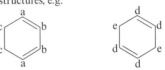

cyclohexa-1,4-diene   cyclohexa-1,3-diene

(The number of chemical environments in cyclohexa-1,3-diene is) three chemical environments and in cyclohexa-1,4-diene it is two. **(1)**

149. Proton NMR spectroscopy

1 B **(1)**

2 Three from the following for **1** mark each:
  • ^{1}H nuclei are highly shielded / signal up field of most other compounds
  • 12 equivalent hydrogen atoms / 4 equivalent carbon atoms / single intense peak
  • unreactive / non-toxic
  • low boiling point (so easily removed)

3 (a) CCl_4 does not contain hydrogen atoms. **(1)**
  (b) $CDCl_3$ contains ^{2}H not ^{1}H. **(1)**

4 (a) There are two different chemical environments. **(1)**
  (b) The area is proportional to the number of hydrogen atoms in each chemical environment **(1)** the peak at 4 ppm represents H in OH / the peak at 3.4 ppm represents H in CH_3. **(1)**

5 (a) 4 peaks **(1)** (b) 3 : 2 : 1 : 3 **(1)**
    CH_3 (on left), CH_2, CHBr, CH_3 (on right) **(1)**

150. Splitting patterns

1 A **(1)**

2 (a) It is adjacent to a CH_2 group **(1)** (using $n + 1$ rule) peak split into $(2 + 1) = 3$ **(1)**.
  (b) It is adjacent to a CHCl group **(1)** (using $n + 1$ rule) peak split into $(1 + 1) = 2$ **(1)**.
  (c) It is split into five / quintet **(1)**, it is adjacent to CH_3 group and CHCl group **(1)** (using $n + 1$ rule) peak split into $(3 + 1) + 1 = 5$ **(1)**.

3 (a) (i) CH_3 adjacent to CH_2 **(1)**
    (ii) CH_3 (not adjacent to non-equivalent hydrogen atoms) **(1)**
    (iii) CH_2 adjacent to CH_3 **(1)**
  (b) Displayed formula (for ethyl methyl ether) **(1)**

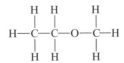

151. AS Practice paper 1

1 (a) C **(1)**
  (b) atoms with the same number of protons **(1)** but different numbers of neutrons **(1)**

(c) 0.07% ^{40}K *and* correct calculation of A_r with working shown, e.g. [(93.2 × 39) + (0.07 × 40) + (6.73 × 41)]/100 = 39.1 **(1)** Correct answer given to three significant figures **(1)**

(d) amount of Na$^+$ = 6.00/58.5 = 0.1026 mol **(1)**
number of ions = 0.1026 × 6.02 × 10^{23} = 6.18 × 10^{22} **(1)**

2 (a) Chlorine atoms have more protons / greater nuclear charge **(1)** smaller atomic radius **(1)**.

(b) The molecules are symmetrical / linear and trigonal planar **(1)** bond polarities cancel out **(1)**.

(c) Diagram with two water molecules with lone pair of electrons on O of one molecule **(1)**
δ^+ on H atom, δ^- on O atom, dashed line from lone pair to H with 180° bond angle H−O−H **(1)**

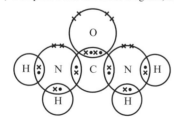

3 (a) **B (1)**

(b) (i) Add ammonia solution **(1)** AgCl(s) dissolves **(1)**
AgI(s) does not dissolve / stays as a yellow solid **(1)**.

(ii) Add silver nitrate solution **(1)** no change with HNO$_3$(aq) **(1)** HBr(aq) gives a cream precipitate **(1)**.

(c) Ba^{2+}(aq) + SO$_4$$^{2-}$(aq) → BaSO$_4$(s)
1 mark for correct species and balanced, **1** mark for correct state symbols

4 (a) **B (1)**   (b) **A (1)**

(c) To kill / prevent growth of: microorganisms / bacteria. **(1)**

(d) **D (1)**

(e) Oxygen is simultaneously oxidised and reduced **(1)** −1 to −2 (reduction) and −1 to 0 (oxidation) **(1)**.

(f) molar mass of PbCl$_2$ = 278.2 **(1)**
amount of PbCl$_2$ = 2.46 / 278.2 = 8.843 × 10^{-3} mol **(1)**
mass of Pb = 207.2 × 8.843 × 10^{-3} = 1.832 g **(1)**
percentage of Pb = 100 × 1.832 / 2.00 = 91.6% **(1)**

5 (a) Suitable method for **1** mark, e.g. time taken for limewater to turn milky or cloudy / time taken for precipitate to first appear

(b) Two variables to control for **1** mark each, e.g.
• mass or amount of group 2 carbonate used
• volume of limewater used
• particle size of group 2 carbonates
• distance of flame to the bottom of the boiling tube
• Bunsen burner flame

(c) Stability increases down Group 2. **(1)**
Ionic radius of Group 2 metal ion increases. **(1)**
Polarising power of metal ion decreases. **(1)**
Carbonate ion becomes less polarised / distorted. **(1)**

(d) Li$_2$CO$_3$ → Li$_2$O + CO$_2$ **(1)**

6 (a) **C (1)**   (b) COCl$_2$ + 4NH$_3$ → (NH$_2$)$_2$CO + 2NH$_4$Cl **(1)**

(c) % by mass of O = 26.66% **(1)**
Calculation of empirical formula, e.g.

N	H	C	O
46.67/14.0	6.67/1.0	20.00/12.0	26.66/16.0 **(1)**
= 3.33	= 6.67	= 1.66	= 1.66
2	4	1	1 **(1)**

(d) Completed dot and cross diagram, e.g.

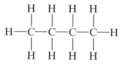

Correct electron pairs for C and O, including lone pairs on O **(1)**
Correct electron pairs for N and H, including lone pair on N atoms **(1)**

(e) Shape is trigonal planar. **(1)**
C=O is treated as a single bond, with shape based on three bond pairs around carbon atom. **(1)**
Electron pairs repel to positions with minimum repulsion / maximum separation. **(1)**

7 (a) **A (1)**

(b) (Going down the group) fluorine and chlorine are gases, bromine is liquid, and iodine is solid **(1)**
Melting temperature *and* boiling temperature increase down the group **(1)**
Number of electrons in each molecule increases **(1)**
Strength of London forces increases **(1)**
More energy is needed to overcome these intermolecular forces **(1)**

(c) (i) H$_2$SO$_4$ + 8H$^+$ + 8e$^-$ → H$_2$S + 4H$_2$O **(1)**
(ii) 8I$^-$ + H$_2$SO$_4$ + 8H$^+$ → 4I$_2$ + H$_2$S + 4H$_2$O **(1)**
(iii) reducing agent **(1)**

(d) (i) hydrogen chloride **(1)**
(ii) orange-brown fumes **(1)**
KBr + H$_2$SO$_4$ → KHSO$_4$ + HBr **(1)**
2HBr + H$_2$SO$_4$ → Br$_2$ + SO$_2$ + 2H$_2$O **(1)**
(iii) reducing ability increases from Cl$^-$ to I$^-$ **(1)**

8 (a) All titres calculated: 23.75, 23.25, 23.50, 23.15, 23.20 cm^3 **(1)**
Runs 2, 4 and 5 ticked *and* mean titre calculated: 23.20 cm^3 **(1)**

(b) (i) amount of NaOH = 25.00 × 10^{-3} × 0.180
= 4.50 × 10^{-3} mol **(1)**
concentration of sulfamic acid
= (4.50 × 10^{-3})/(23.20 × 10^{-3}) = 0.194 mol dm^{-3} **(1)**

(ii) molar mass of sulfamic acid = 97.1
concentration of sulfamic acid = 97.1 × 0.194 = 18.8 g dm^{-3} **(1)**

(iii) mass in 250 cm^3 = 18.80 × 250 × 10^{-3} = 4.7 g **(1)**
percentage purity = 100 × 4.7/4.8 = 97.9% **(1)**

(c) (i) balance (two readings) = 2 × 100 × 0.05/4.8 = 2.08%
volumetric pipette (one reading) = 100 × 0.04/25.00 = 0.160%
*Both needed for **1** mark*

(ii) total error = (2.08 + 0.200 + 0.160 + 0.413 + 0.413) = 2.85% **(1)**

(iii) range is 95.1 − 100.7 % / 97.9 ± 2.85% of this value so the claim could be correct **(1)**

(d) Three sources of error for **1** mark each, with **1** mark for suitable improvement for that error, e.g.
• balance largest percentage error − use a 2 d.p. balance/more precise balance
• acid left in the weighing boat − weigh by difference / rinse and add washings to beaker
• acid left in the beaker − rinse and add washings to the volumetric flask
• contents of volumetric flask not mixed − invert flask to mix contents
• burette / volumetric pipette not rinsed with reagent − rinse with acid / sodium hydroxide

154. AS Practice paper 2

1 (a) **B (1)**

(b) (i) (2436.98 − 2436.52) = 0.46 g **(1)**
(ii) rearrange ideal gas equation: $n = pV/RT$ **(1)**
n = (101 325 × 238 × 10^{-6})/(8.31 × 293) **(1)**
n = 9.90 × 10^{-3} mol **(1)**
(iii) molar mass = 0.46/(9.90 × 10^{-3}) = 46.5 mol^{-1} **(1)**
answer to 1 d.p.

(c) (i) Displayed formula of butane **(1)**

$$\text{H}-\overset{\overset{\displaystyle \text{H}}{|}}{\underset{\underset{\displaystyle \text{H}}{|}}{\text{C}}}-\overset{\overset{\displaystyle \text{H}}{|}}{\underset{\underset{\displaystyle \text{H}}{|}}{\text{C}}}-\overset{\overset{\displaystyle \text{H}}{|}}{\underset{\underset{\displaystyle \text{H}}{|}}{\text{C}}}-\overset{\overset{\displaystyle \text{H}}{|}}{\underset{\underset{\displaystyle \text{H}}{|}}{\text{C}}}-\text{H}$$

(ii) C$_2$H$_5$ **(1)**

2 (a) **C (1)**

(b) Pressure **(1)**

(c) **B (1)**

(d) enthalpy change when one mole of water is produced by the neutralisation of an acid with an alkali **(1)** measured at 100 kPa with a specified temperature usually 298 K **(1)** with all solutions at 1 mol dm^{-3} **(1)**

3 (a) Answer should include the following points in a logical sequence for **1** mark each:
- forward reaction is exothermic
- so reducing the temperature will increase the equilibrium yield of sulfur trioxide
- but reducing the temperature decreases the rate of reaction
- because fewer reactant molecules have the activation energy or more
- so the rate of successful collisions will decrease
- increased yield and decreased rate oppose each other so overall effect difficult to predict

(b) (i) Curve starting at reactants level and finishing at product level, with height of peak higher than the original curve **(1)**

(ii) $\Delta_r H$ from reactants level to product level **(1)** E_a from reactants level to peak of catalysed reaction **(1)**

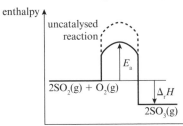

(c) Correct expression for K_c for **1** mark,

e.g. $K_c = \dfrac{[SO_3(g)]^2}{[SO_2(g)]^2[O_2(g)]}$

4 (a) (i) Vertical line drawn at the peak of the printed curve and labelled **A (1)**

(ii) Curve starts at the origin, and is displaced to the right with a lower peak the original **(1)**
Curve crosses the printed curve once only, and afterwards does not touch the printed curve **(1)**

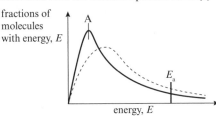

(b) (i) Vertical line drawn to the left of the printed E_a line and labelled **B (1)**

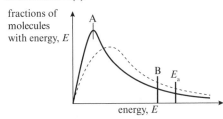

(ii) Answer should include the following for **1** mark each:
- area under the curve to the right of the activation energy represents the number of molecules with the activation energy or more
- area under the curve is greater with the catalyst than without it
- the rate of successful collisions is greater with the catalyst

5 (a) (i) $(243.52 - 242.64) = 0.88$ g **(1)**

(ii) $(34.3 - 19.8) = 14.5\,°C$ **(1)**

(b) Using $Q = mc\Delta T$
Energy change of water $= 200.0 × 4.18 × 14.5 = 12122$ J **(1)**
$= 12.122$ kJ **(1)**
Amount of ethanol burned $= 0.88/46.0 = 0.019\,13$ mol **(1)**
Enthalpy change of combustion $= -12.122/0.019\,13 = -634\,\text{kJ mol}^{-1}$ **(1)** *answer must be negative and to three significant figures for the mark*

(c) Two reasons from the following for **1** mark each:
- incomplete combustion occurred
- the apparatus / surroundings were also heated
- the conditions were not standard conditions

(d) (i) energy transferred to reactants to break bonds:
$(1 × C–C) + (5 × C–H) + (1 × C–O) + (1 × O–H) + (3 × O=O)$
$(1 × 347) + (5 × 413) + (1 × 358) + (1 × 464) + (3 × 498)$
$347 + 2065 + 358 + 464 + 1494 = 4728\,\text{kJ mol}^{-1}$ **(1)**
energy transferred to surroundings when bonds made:
$(4 × C=O) + (6 × O–H)$
$(4 × 805) + (6 × 464)$
$3220 + 2784 = 6004\,\text{kJ mol}^{-1}$ **(1)**
enthalpy change $= 4728 - 6004 = -1276\,\text{kJ mol}^{-1}$ **(1)**

(ii) Actual bond enthalpies in the reactants and products may differ from the mean values used **(1)** mean bond enthalpies are for substances in the gas state but ethanol will be in the liquid state **(1)**.

6 (a) **D (1)**

(b) ultraviolet light / sunlight **(1)**

(c) (i) $Cl_2 \rightarrow 2{\cdot}Cl$ **(1)**

(ii) $C_2H_6 + {\cdot}Cl \rightarrow {\cdot}CH_2CH_3 + HCl$ **(1)**
${\cdot}CH_2CH_3 + Cl_2 \rightarrow CH_3CH_2Cl + {\cdot}Cl$ **(1)**

(iii) ${\cdot}CH_2CH_3 + {\cdot}Cl \rightarrow CH_3CH_2Cl$ **(1)**

(d) (i) Two ${\cdot}CH_2CH_3$ radicals combine **(1)**
$2{\cdot}CH_2CH_3 \rightarrow C_4H_{10}$ **(1)**

(ii) fractional distillation **(1)**

7 (a) (i) $CH_3CH_2CH_2CH=CH_2 + Br_2 \rightarrow$
$CH_3CH_2CH_2CHBrCH_2Br$ **(1)**
1,2-dibromopentane **(1)**

(ii) brown / red-brown to colourless **(1)**

(b) (i) Labelled displayed formulae of E/Z isomers of pent-2-ene for **1** mark each, e.g.

E-pent-2-ene Z-pent-2-ene

(ii) The carbon atoms involved in the C=C bond in pent-2-ene both have different groups attached **(1)** but in pent-1-ene one of the carbon atoms involved in the C=C bond has two identical groups attached / two H atoms attached **(1)**.

(c) (i) The molecule contains a C=C bond / carbon–carbon double bonds. **(1)**

(ii) One of these isomers for **1** mark:

(d) (i) electrophilic addition **(1)**

(ii) reaction mechanism, e.g.

- curly arrow from C=C bond to δ^+ H atom *and* curly arrow from H−Br bond to δ^- bromine atom **(1)**
- correct structure of secondary carbocation with + charge shown **(1)**
- curly arrow from Br^- to C+ with correct structure of product **(1)**

8 (a) **1** mark for each correct skeletal formula, e.g.

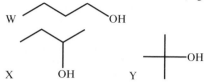

(b) **1** mark for each of the following, written in a logical order:
- acidify potassium dichromate(VI) with sulfuric acid and add to each alcohol
- warm the mixture under distillation conditions
- colour change (from orange to green) with W and X
- no change with Y
- add organic product from W and X to Fehling's solution and warm
- orange / red precipitate forms with product from W but no colour change / stays blue with product from X

(c) (i) +1 **(1)**

(ii) $m/z = 43$ is due to $(CH_3)_2CH^+ / CH_3CHCH_3^+$ **(1)**

$m/z = 31$ is due to CH_2OH^+ **(1)**

(iii) primary **(1)**

157. A2 Practice paper 1

1 (a) **D (1)**　　　(b) **C (1)**　　　(c) **D (1)**

(d) (i) (trigonal) pyramidal **(1)**

(ii) 107° **(1)** three bond pairs and one lone pair of electrons on P atom **(1)** lone pair−bond pair repulsion is greater than bond pair−bond pair repulsion **(1)**

2 (a) **A (1)**　　　(b) **C (1)**

(c) enthalpy change **(1)** to form a mole of gaseous ions with a single positive charge / to remove one mole of electrons **(1)** from (a mole) of gaseous atoms **(1)**

(d) Balanced cycle **(1)**

substitution / rearrangement for E_{aff} **(1)**

Correct answer with correct sign **(1)**

For example:

$(2 \times 107) + 249 + (2 \times 496) + (-141) + E_{aff,2} + (-2478)$
$= -414$ **(1)**

$E_{aff,2} = -414 - 214 - 249 - 992 + 141 + 2478$ **(1)**

$E_{aff,2} = +750 \text{ kJ mol}^{-1}$ **(1)**

3 (a) (i) Mn　　$1s^2 2s^2 2p^6 3s^2 3p^6 3d^5 4s^2$ **(1)**

(ii) Mn²⁺　　$1s^2 2s^2 2p^6 3s^2 3p^6 3d^5$ **(1)**

(b) (i) $VO_2^+(aq)$ **(1)**

Reaction is feasible if E_{cell} is positive **(1)**

E_{cell} is positive for reaction of MnO_4^- with V^{2+} to form V^{3+}, for V^{3+} to form VO^{2+}, and for VO^{2+} to form VO_2^+ **(1)**

(ii) $V^{2+}(aq) + 2H_2O(l) \rightarrow VO_2^+(aq) + 4H^+(aq) + 3e^-$ **(1)**

(iii) $5V^{2+}(aq) + 3MnO_4^- + 4H^+(aq) \rightarrow 5VO_2^+(aq) + 3Mn^{2+} + 2H_2O(l)$ **(1)**

4 (a) 1,2-diaminoethane has two lone pairs of electrons **(1)** so it can form two dative covalent bonds with a central metal ion **(1)**.

(b) (i) Graph plotted, e.g.

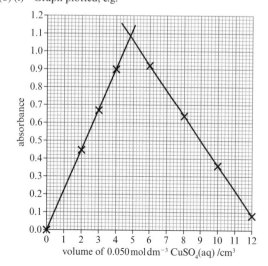

1 mark for labelled axes the correct way round, scale chosen so that plotted points occupy more than half the graph paper on each axis

1 mark for all points plotted to ±1 mm, two lines as required, intersect at about 4.8, 1.08

(ii) 4.8 cm³ **(1)**

(iii) $(12.0 - 4.8) = 7.2 \text{ cm}^3$ **(1)**

(c) amount of $Cu^{2+} = 4.8 \times 10^{-3} \times 0.050 = 2.40 \times 10^{-4} \text{ mol}$ **(1)**

amount of $H_2NCH_2CH_2NH_2 = 7.2 \times 10^{-3} \times 0.100 = 7.20 \times 10^{-4} \text{ mol}$ **(1)**

(d) (i) ratio $= (7.20 \times 10^{-4})/(2.40 \times 10^{-4}) = 3$ **(1)**

(ii) formula is $[Cu(en)3]^{2+}/[Cu(H_2NCH_2CH_2NH_2)_3]^{2+}$ **(1)**

5 (a) (i) bleach **(1)**

(ii) The oxidation number of Cl in NaClO is +1 **(1)**. It changes to −1 in NaCl *and* to +5 in NaClO₃ **(1)**. Chlorine has been both reduced and oxidised **(1)**.

(b) (i) The oxidation number of chlorine (and the other elements present) stay the same **(1)** so it is not a redox reaction **(1)**.

(ii) The hydrogen chloride dissolves in water vapour in the air **(1)** forming droplets of hydrochloric acid **(1)**.

(c) (i) $H_2SO_4 + 2H^+ + 2e^- \rightarrow SO_2 + 2H_2O$ **(1)**

(ii) $H_2SO_4 + 2H^+ + 2I^- \rightarrow SO_2 + 2H_2O + I_2$ **(1)**

(iii) reducing agent **(1)**

6 (a) amount $= 345/24.0 = 14.375 \text{ mol}$ **(1)**

(b) amount of CO_2 produced $= 14.375 \times 3 \times 8 = 345 \text{ mol}$ **(1)**

amount of LiOH needed $= 2 \times 345 = 690 \text{ mol}$ **(1)**

molar mass of LiOH $= (6.9 + 16.0 + 1.0) = 23.9 \text{ g mol}^{-1}$

mass of LiOH $= 690 \times 23.9 = 16 491 \text{ g}$ **(1)**

mass of LiOH $= 16.5 \text{ kg}$ **(1)** *must be in kg to three significant figures for this mark*

(c) molar mass of $Li_2O_2 = (2 \times 6.9) + (2 \times 16.0) = 45.8 \text{ g mol}^{-1}$ **(1)**

mass of $Li_2O_2 = 345 \times 45.8 = 15 801 \text{ g} = 15.8 \text{ kg}$ **(1)**

(d) lower mass of lithium peroxide needed **(1)** oxygen is released in the reaction **(1)**

7 (a) **C (1)**

(b) Correct expression for K_c for **1** mark:

$$K_c = \frac{[CO_2(g)][H_2(g)]^4}{[CH_4(g)][H_2O(g)]^2}$$

(c) (i) At equilibrium:

amount of $CH_4 = 0.80 - 0.20 = 0.60 \text{ mol}$ **(1)**

amount of $H_2O = 1.6 - (2 \times 0.20) = 1.2 \text{ mol}$ **(1)**

amount of $H_2 = 4 \times 0.20 = 0.80 \text{ mol}$ **(1)**

(ii) All four concentrations for **1** mark:

$[CH_4(g)] = 0.60/2.5 = 0.24 \text{ mol dm}^{-3}$

$[H_2O(g)] = 1.2/2.5 = 0.48 \text{ mol dm}^{-3}$

$[CO_2(g)] = 0.20/2.5 = 0.08 \text{ mol dm}^{-3}$

$[H_2(g)] = 0.80/2.5 = 0.32 \text{ mol dm}^{-3}$

(iii) Correct substitution for **1** mark, e.g.

$$K_c = \frac{0.08 \times (0.32)^4}{0.24 \times (0.48)^2}$$

$= 0.015$ **(1)**

$\text{mol}^2 \text{ dm}^{-6}$ **(1)**

8 (a) **B (1)**

(b) Ethanoic acid is only partially ionised in solution. **(1)**

(c) $[H^+] = 10^{-pH} = 0.50 \text{ mol dm}^{-3}$ **(1)**

(d) (i) solution which resists changes in pH **(1)** when small amounts of acid or base are added **(1)**

(ii) $CH_3COO^-(aq) + H^+(aq) \rightarrow CH_3COOH(aq) /$
$CH_3COONa(aq) + HCl(aq) \rightarrow CH_3COOH(aq) + NaCl(aq)$ **(1)**

(iii) $[CH_3COOH] = (125 \times 10^{-3} \times 0.250)/(200 \times 10^{-3})$
$= 0.15625 \text{ mol dm}^{-3}$ *and*
$[CH_3COO^-] = (75 \times 10^{-3} \times 0.100)/(200 \times 10^{-3})$
$= 0.0375 \text{ mol dm}^{-3}$ **(1)**

Expression for **1** mark, e.g.

$$[H^+] = K_a \times \frac{[CH_3COOH]}{[CH_3COO^-]}$$

$[H+] = 1.74 \times 10^{-5} \times 0.15625/0.0375$
$= 7.25 \times 10^{-5} \text{ mol dm}^{-3}$ **(1)**

$pH = -\log_{10}(7.25 \times 10^{-5}) = 4.14$ **(1)**

9 (a) (i) **C (1)**
　　 (ii) **A (1)**
　(b) (i) $[Co(H_2O)_6]^{2+}$ **(1)**
　　 (ii) Octahedral **(1)**
　　 (iii) $[Co(H_2O)_6]^{2+} + 6NH_3 \rightarrow [Co(NH_3)6]^{2+} + 6H_2O$ **(1)**
　　　　 ligand exchange **(1)**
10 (a) **D (1)**
　(b) (As the group is descended) they become more stable **(1)**
　　 the ionic radius increases / charge density of metal ions decreases **(1)**
　　 the nitrate ions are less polarised **(1)**
　(c) $\Delta H_{solution} = -351 + (-364) - (-711)$ **(1)**
　　 $= -4\,kJ\,mol^{-1}$ **(1)** *correct sign needed for the mark*
　(d) (i) $\Delta_r H^\ominus = -635 + (-394) - (-1207) = +178\,kJ\,mol^{-1}$
　　　　 (1) *correct sign needed for the mark*
　　 (ii) $\Delta S^\ominus_{system} = 40 + 214 - 93 = +161\,J\,K^{-1}\,mol^{-1}$ **(1)**
　　　　 correct sign needed for the mark
　　 (iii) The system has become more disordered **(1).**
　　 (iv) $\Delta G = \Delta H - T\Delta S^\ominus_{system}$
　　　　 $\Delta G = +178 - (298 \times 161 \times 10^{-3}) = +130\,kJ\,mol^{-1}$ **(1)**
　　　　 correct sign needed for the mark
　　　　 ΔG is positive so the reaction cannot be feasible **(1).**
　　 (v) (For $\Delta G = 0$) $T = \Delta H/\Delta S_{system} = 178/(161 \times 10^{-3})$ **(1)**
　　　　 $T = 1106\,K$ **(1)**

160. A2 Practice paper 2

1 (a) **A (1)**
　(b) **C (1)**
　(c) overlap between p-orbitals of the two carbon atoms to form a sigma bond / σ bond **(1)**
　　 remaining p-orbitals overlap to form pi bond / a π bond **(1)**
　　 axial overlap / end-on overlap / head on overlap for σ bond *and* sideways overlap for π bond **(1)**
　　 These marks can be gained from labelled diagram / diagrams
2 (a) **D (1)**
　(b) **A (1)**
　(c) **C (1)**
3 (a) proton acceptor **(1)**
　(b) **C (1)**
　(c) (i) $CH_3CH_2NH_3^+NO_3^-$ **(1)**
　　 (ii) $(CH_3CH_2NH_3^+)_2SO_4^{2-}$ / $CH_3CH_2NH_3^+HSO_4^-$ **(1)**
　(d) (i) Butylamine forms hydrogen bonds with water **(1)** via the NH_2 group **(1).**
　　 (ii) $CH_3CH_2CH_2CH_2NH_2 + H_2O \rightleftharpoons$
　　　　 $CH_3CH_2CH_2CH_2NH_3^+ + OH^-$ **(1)**
　　　　 $\rightarrow$ *instead of* $\rightleftharpoons$ *is acceptable*
4 (a) glycine: aminoethanoic acid **(1)**
　　 alanine: 2-aminopropanoic acid **(1)**
　(b) alanine has optical isomers but glycine does not **(1)**
　　 the carbon atom attached to the NH_2 group is a chiral centre / has four different groups attached **(1)**
　(c) (i) $^+NH_3CH_2COOH$ **(1)**
　　 (ii) $NH_2CH_2COO^-$ **(1)**
　(d) *Answer must form a logical sequence to gain full marks:*
　　 Alanine, butanoic acid and hexane have similar molar masses **(1)**
　　 so they will have similar numbers of electrons and similar London forces between their molecules **(1)**
　　 but hexane has the lowest melting temperature because it only has London forces between its molecules **(1)**
　　 butanoic acid has a higher melting temperature because it also has hydrogen bonds between its molecules **(1)**
　　 alanine has the highest melting temperature because it exists as zwitterions **(1)**
　　 with strong ionic bonds between these zwitterions **(1).**
5 (a) **C (1)**
　(b) **C (1)**
　(c) Compound A is unsaturated but compound B is saturated **(1)** compound B could be a cyclic compound / named six-carbon saturated cyclic compound, e.g. cyclohexane, methylcyclopentane **(1).**

(d) (i) Reaction mechanism, e.g.

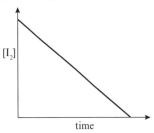

　　 • curly arrow from C=C bond to δ^+ H atom *and* curly arrow from H−Br bond to δ^- bromine atom **(1)**
　　 • correct structure of primary carbocation with + charge shown **(1)**
　　 • curly arrow from Br to C+ with correct structure of product **(1)**
　 (ii) 2-bromobutane **(1)** forms via a secondary carbocation **(1)** which is more stable than the primary carbocation which leads to the formation of the minor product **(1)**

6 (a) **B (1)**
　(b) the power of the concentration term for the reactant **(1)** in the rate equation **(1)**
　(c) (i) 2 / second order **(1)**
　　 (ii) catalyst **(1)** because they appear in the rate equation but not in the balanced equation **(1)**
　　 (iii) **1** mark for a graph of $[I_2]$ on the vertical axis, time on the horizontal axis, straight line with negative gradient, e.g.

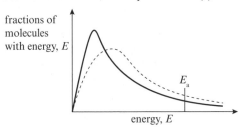

　　 (iv) rate increases if concentration of propanone increases **(1)**
　　　　 because the rate of collisions increases / collisions are more frequent **(1)**
　(d) (i) 2 / second order **(1)**
　　 (ii) First order
　　　　 comparing runs 1 and 3 [H_2] and [NO] double, and rate increases 8 times **(1)**
　　　　 rate would increase 4 times due to [NO] change (so remainder is 2 times when [H_2] doubles **(1)**
　　 (iii) rate= $k[H_2][NO]^2$ **(1)**
　　 (iv) **1** mark for correct substitution in run 1, 2 or 3, with correct answer, e.g.
　　　　 $k = (1.08 \times 10^{-5})/[(0.85 \times 10^{-3})(2.6 \times 10^{-3})^2] = 1880$ **(1)**
　　　　 units: $mol^{-2}\,dm^6\,s^{-1}$ **(1)**
　(e) (i) Curve starts at the origin, and is displaced to the right with a lower peak than the original **(1)**
　　　　 Curve crosses the printed curve once only, and afterwards does not touch the printed curve **(1).**

(ii) At a higher temperature a greater proportion / number / fraction of molecules have $\geqslant$ activation energy E_a **(1)** so a greater frequency of successful collisions **(1)**.

7 (a) (i) **concentrated** sulfuric acid **(1)** and **concentrated** nitric acid **(1)**
$H_2SO_4 + HNO_3 \rightarrow HSO_4^- + H_2O + NO_2^+$ /
$2H_2SO_4 + HNO_3 \rightarrow 2HSO_4^- + H_3O^+ + NO_2^+$ **(1)**

(ii) Reaction mechanism:

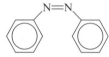

1 mark for arrow from circle to $^+NO_2$
1 mark for intermediate (part circle no further than limits shown)
1 mark for arrow from C–H bond

(b) (i) tin / Sn and concentrated hydrochloric acid **(1)**
(ii) $C_6H_5NO_2 + 6[H] \rightarrow C_6H_5NH_2 + 2H_2O$ **(1)**

(c) (i) $m/z = 182$: molecular ion **(1)**
$m/z = 105$: $C_6H_5NN^+$ **(1)** $m/z = 77$: $C_6H_5^+$ **(1)**

(ii) Diagram of *cis*-azobenzene for **1** mark

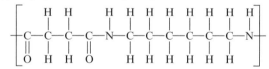

There is a (nitrogen–nitrogen) double bond with restricted movement **(1)** so the two benzene rings can occupy two different positions in space **(1)**.

8 (a) (i) CH_2=CHCl **(1)**
(ii) Displayed formula of repeating unit for poly(chloroethene) for **1** mark

$$\left[\begin{array}{cc} \text{H} & \text{H} \\ | & | \\ \text{C} - \text{C} \\ | & | \\ \text{H} & \text{Cl} \end{array} \right]$$

(iii) 100% **(1)** because there is only one product **(1)**

(b) (i) ethane-1,2-diol **(1)**
(ii) condensation polymerisation **(1)**
(iii) alkaline hydrolysis **(1)** occurs involving OH^- ions **(1)** and δ^+ carbon atoms in C=O bonds **(1)**

(c) Displayed formula of repeating unit for **1** mark

$$\left[\begin{array}{c} \text{H H} \quad \text{H H H H H H H H} \\ | \; | \quad | \; | \; | \; | \; | \; | \; | \; | \\ \text{C–C–C–C–N–C–C–C–C–C–C–N} \\ || \; | \quad || \quad | \; | \; | \; | \; | \; | \\ \text{O H H O} \quad \text{H H H H H H} \end{array} \right]$$

9 (a) (i) oxidation / redox **(1)**
(ii) potassium dichromate(VI) / $K_2Cr_2O_7$ **(1)** acidified with dilute sulfuric acid **(1)** heated under distillation conditions **(1)**

(b) (i) nucleophilic addition **(1)**
(ii) hydrogen cyanide / HCN **(1)** in the presence of potassium cyanide / KCN **(1)**

(c) The C=O group is planar **(1)** so the nucleophile / CN^- ion can attack equally from above or below **(1)** producing a racemic mixture / mixture of two optical isomers in equal amounts **(1)**.

163. A2 Practice paper 3

1 (a) (i) AgCl white **(1)**, AgBr cream **(1)**, AgI yellow **(1)**
(ii) silver fluoride is soluble / does not form a precipitate **(1)**
(iii) nitric acid **(1)** prevents other precipitates (e.g. Ag_2CO_3) forming in the test **(1)**

(b) Add dilute ammonia solution to each precipitate the chloride precipitate redissolves **(1)** but the bromide and iodide precipitates do not **(1)**.
Add concentrated ammonia solution to each precipitate the chloride and bromide precipitates redissolve **(1)** but iodide precipitate does not **(1)**.

(c) *Answer must form a logical sequence to gain full marks:*
• add water to dissolve the mixture **(1)**
• add silver nitrate solution **(1)**
• forming precipitates (of AgCl and AgI) **(1)**
• add excess concentrated ammonia solution **(1)**
• filter the mixture **(1)**
• wash the filtrate with water and dry **(1)**

2 (a) Cu^{2+} / $[Cu(H_2O)_6]^{2+}$ **(1)** Co^{2+} **(1)**
(b) (i) C is $[Co(H_2O)_4(OH)_2]$ **(1)**
D is $[Co(NH_3)_6]^{2+}$ **(1)**
E is $[Co(NH_3)_6]^{3+}$ **(1)**
D darkens because cobalt(II) is oxidised by oxygen in the air to cobalt(III) **(1)**

(ii) F is $[CoCl_4]^{2-}$ **(1)**
$[Co(H_2O)_6]^{2+} + 4Cl^- \rightleftharpoons [CoCl_4]^{2-} + 6H_2O$ **(1)**

(iii) $Ba^{2+}(aq) + SO_4^{2-}(aq) \rightarrow BaSO_4(s)$
1 mark for equation, **1** mark for correct state symbols

(c) Diagram with 3D shape correctly shown **(1)** six electron pairs / six Co–O bonds **(1)**, e.g.

$$\begin{array}{c} \text{OH}_2 \\ \text{H}_2\text{O} \quad | \quad \text{OH}_2 \\ \diagdown \; \text{Co} \; \diagup \\ \text{H}_2\text{O} \diagup \quad | \quad \diagdown \text{OH}_2 \\ \text{OH}_2 \end{array}$$

90° **(1)** octahedral **(1)**

(d) $CoSO_4$ **(1)**

3 (a) D **(1)**   (b) 2-methylpropan-2-ol **(1)**
(c) Order with respect to $(CH_3)_3CBr$ is first order **(1)** comparing experiments 1 and 2, $[(CH_3)_3CBr]$ doubles and so does the initial rate **(1)**
Order with respect to OH^- ions is zero order **(1)** comparing experiments 2 and 3, $[OH^-]$ doubles and but the initial rate stays the same **(1)**

(d) rate $=k[(CH_3)_3CBr]$ / rate $= k[(CH_3)_3CBr][OH^-]^0$ **(1)**
(e) $k = (2.86 \times 10^{-5})/(1.2 \times 10^{-2}) = 2.4 \times 10^{-3}$ **(1)** units: s^{-1} **(1)**
Answer must be to two significant figures for the first mark

(f) (i) Step 1 **(1)** because the rate-determining step is the slowest one and involves $(CH_3)_3CBr$ / step 2 involves OH^- which is not in the rate equation so this must be the fastest step **(1)**

(ii) Reaction mechanism, e.g.

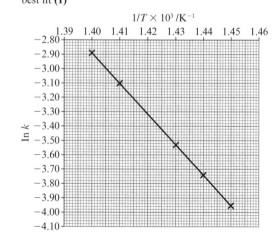

• curly arrow from C–Br bond to Br **(1)**
• correct structure of tertiary carbocation with + charge shown *and* curly arrow from OH^- to C+ with correct structure of product **(1)**

(iii) S_N1 because there is only one reactant particle in the rate-determining step **(1)**

4 (a) Axes labelled and scale chosen so that plotted points occupy >50% of each axis **(1)**
Points plotted correct to ± 1 mm with suitable line of best fit **(1)**

(b) Gradient calculated with working shown for **1** mark, e.g.
$[-3.95 -(-2.89)]/[(1.45 \times 10^{-3}) - (1.40 \times 10^{-3})]$
$= -1.06/(0.05 \times 10^{-3}) = -21\,200$
1 mark for negative gradient

(c) $E_a = -$gradient $\times R/1000$
$= 21\,200 \times 8.31/1000 = 176$ **(1)** kJ mol^{-1} **(1)**
Answer must be to three significant figures for first mark, second mark given if it is consistent with the value for E_a

5 (a) 0 / zero **(1)** because $O_2(g)$ is an element in its standard state **(1)**

(b) (i) $\Delta_r H^{\ominus} = -441 -(-297 + 0) = -144$ kJ mol^{-1} **(1)**

(ii) The data book quotes the value when SO_3 is in the gas state but the table shows it in the liquid state **(1)** the difference will be the energy required / enthalpy change to change SO_3 from a liquid to a gas **(1)**.

(c) $\Delta S^{\ominus} = 96 - (248 + 205/2) = -254.5$ J K^{-1} mol^{-1} **(1)**

(d) Use of $\Delta G = \Delta H - T\Delta S$
$= -98.5 -(298 \times (-254.5/1000))$ **(1)** $= -22.7$ kJ mol^{-1} **(1)**
ΔG is negative at this temperature so the reaction is feasible. **(1)**

(e) As the temperature increases $T\Delta S$ becomes more negative **(1)** so at very high temperatures $\Delta H - T\Delta S$ becomes positive **(1)**.

(f) (i) The position of equilibrium moves to the left **(1)** in the direction of the endothermic change **(1)**.

(ii) The position of equilibrium moves to right **(1)** as there are fewer moles of gas on the right / to maintain value of K_p **(1)**.

6 (a) M_r of aspirin $= (9 \times 12.0) + (4 \times 16.0) + (8 \times 1.0)$
$= 108.0 + 64.0 + 8.0 = 180.0$ *and* M_r of ethanoic acid
$= (2 \times 12.0) + (2 \times 16.0) + (4 \times 1.0) = 24.0 + 32.0 + 4.0$
$= 60.0$ **(1)**
Atom economy $= (100 \times 180.0)/(180.0 + 60.0) = 75\%$ **(1)**

(b) Dissolve in a **minimum** volume of hot water **(1)** filter while hot **(1)** cool and filter while cold **(1)** wash (on filter) with cold water **(1)**.

(c) (i) amount of NaOH $= 50.0 \times 10^{-3} \times 1.00$
$= 5.00 \times 10^{-2}$ mol **(1)**

(ii) amount of HCl $= 23.60 \times 10^{-3} \times 0.150$
$= 3.54 \times 10^{-3}$ mol **(1)**
amount of NaOH in each portion
$= 3.54 \times 10^{-3}$ mol **(1)**
total amount of NaOH remaining
$= 10 \times 3.54 \times 10^{-3} = 3.54 \times 10^{-2}$ mol **(1)**

(iii) amount of NaOH used
$= (5.00 \times 10^{-2}) - (3.54 \times 10^{-2})$ mol
$= 1.46 \times 10^{-2}$ mol **(1)**

(iv) amount of aspirin hydrolysed $= 1.46 \times 10^{-2} \div 2$
$= 7.30 \times 10^{-3}$ mol **(1)**

(v) mass of aspirin hydrolysed $= 180.0 \times 7.30 \times 10^{-3}$
$= 1.314$ g **(1)**
percentage purity $= (100 \times 1.314)/1.52 = 86.4\%$ **(1)**
answer must be to three significant figures to gain the mark

7 (a) **B (1)**

(b) (i) Potentials cannot be measured directly **(1)** two electrodes / connections to a voltmeter are needed to measure a potential difference **(1)**.

(ii) Sketch of standard hydrogen electrode labelled to show:
- $H_2(g)$ **(1)**
- platinum electrode/platinum black electrode **(1)**
- $H^+(aq)$/HCl(aq) **(1)**
- 100 kPa **(1)** 1.0 mol dm^{-3} **(1)** 298 K **(1)**

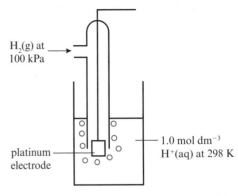

(iii) to allow ions to move between half-cells **(1)** which completes the circuit **(1)**

(c) chloride ions do not react with dichromate(VI) ions but chloride ions do react with manganate(VII) ions **(1)**
$E^{\ominus}_{cell} = 1.51 - 1.36 = 0.15$ V for this reaction **(1)**
chlorine is produced which is toxic / gives an inaccurate titre **(1)**

(d) (i) negative electrode: $\frac{1}{2}H_2 \rightarrow H^+ + e^- / H_2 \rightarrow 2H^+ + 2e^-$ **(1)**
positive electrode: $\frac{1}{2}O_2 + 2H^+ + 2e^- \rightarrow H_2O$ **(1)**

(ii) $E^{\ominus}_{cell} = 1.23 - 0.00 = 1.23$ V **(1)**

(iii) same overall reaction so $E^{\ominus}_{cell}$ must also be 1.23 V **(1)**
$E^{\ominus} = 0.40 - 1.23 = -0.83$ V **(1)**
$H_2 + 2OH^- \rightarrow 2H_2O + 2e^-$ **(1)**

8 (a) Two from the following for **1** mark each: rapid / accurate / sensitive

(b) Ethanoic acid has a different M_r because it contains different numbers of atoms of each element / has a different formula **(1)** the other three have the same M_r because they have the same molecular formula / same number of atoms of each element **(1)**.

(c) $m/z = 31$ $^+CH_2OH$ **(1)**
$m/z = 45$ $^+CH(OH)CH_3$ **(1)**

(d) ethanoic acid *and* propan-2-ol **(1)**

(e) triplet due to CH_3 adjacent to CH_2 so split into $2 + 1 = 3$ **(1)**
quartet due to CH_2 adjacent to CH_3 so split into $3 + 1 = 4$ **(1)**
single due to CH_3 adjacent to O so no splitting **(1)**

(f) propan-2-ol **(1)**
singlet due to H in OH **(1)**
doublet due to two equivalent CH_3 groups adjacent to CH so split into $1 + 1 = 2$ **(1)**
septet due to H in CH group adjacent to a CH_3 group on each side so split into $3 + 3 + 1 = 7$ **(1)**

(g) *Answer must form a logical sequence to gain full marks:*
- Spectrum 1 could be due to propan-1-ol or to propan-2-ol, and Spectrum 2 due to ethanoic acid **(1)**
- in Spectrum 1, absorbance at range 3750−3200 cm^{-1} due to O−H in alcohols **(1)**
- in Spectrum 2, absorbance at range 3300−2500 cm^{-1} due to O−H in carboxylic acids **(1)**
- in Spectrum 2, absorbance at range 1725−1700 cm^{-1} due to C=O **(1)**
- but in Spectrum 1, no significant absorbance at range 1725−1700 cm^{-1} **(1)**
- ethanoic acid has C=O group but propan-1-ol and propan-2-ol do not **(1)**

For your own notes

For your own notes

For your own notes

Published by Pearson Education Limited, 80 Strand, London, WC2R 0RL.

www.pearsonschoolsandfecolleges.co.uk

Copies of official specifications for all Edexcel qualifications may be found on the website: www.edexcel.com

Text and illustrations © Pearson Education Limited 2015
Copyedited by Hilary Herrick and Marilyn Grant
Typeset and illustrations by Tech-Set Ltd, Gateshead
Produced by Out of House Publishing
Cover illustration by Miriam Sturdee

The right of Nigel Saunders to be identified as author of this work has been asserted by him in accordance with the Copyright, Designs and Patents Act 1988.

First published 2016

18 17 16
10 9 8 7 6 5 4 3 2

British Library Cataloguing in Publication Data
A catalogue record for this book is available from the British Library

ISBN 9781447989943

Printed in Slovakia by Neografia

A note from the publisher
In order to ensure that this resource offers high-quality support for the associated Pearson qualification, it has been through a review process by the awarding body. This process confirms that this resource fully covers the teaching and learning content of the specification or part of a specification at which it is aimed. It also confirms that it demonstrates an appropriate balance between the development of subject skills, knowledge and understanding, in addition to preparation for assessment.

Endorsement does not cover any guidance on assessment activities or processes (e.g. practice questions or advice on how to answer assessment questions), included in the resource nor does it prescribe any particular approach to the teaching or delivery of a related course.

While the publishers have made every attempt to ensure that advice on the qualification and its assessment is accurate, the official specification and associated assessment guidance materials are the only authoritative source of information and should always be referred to for definitive guidance.

Pearson examiners have not contributed to any sections in this resource relevant to examination papers for which they have responsibility.

Examiners will not use endorsed resources as a source of material for any assessment set by Pearson.

Endorsement of a resource does not mean that the resource is required to achieve this Pearson qualification, nor does it mean that it is the only suitable material available to support the qualification, and any resource lists produced by the awarding body shall include this and other appropriate resources.